Vector Analysis for Computer Graphics

John Vince

Vector Analysis
for Computer Graphics

Second Edition

 Springer

John Vince
Bournemouth University
Poole, UK

ISBN 978-1-4471-7507-0   ISBN 978-1-4471-7505-6   (eBook)
https://doi.org/10.1007/978-1-4471-7505-6

This Springer imprint is published by the registered company Springer-Verlag London Ltd.
part of Springer Nature.
The registered company address is: The Campus, 4 Crinan Street, London, N1 9XW, United Kingdom

In the hope that some of my grandchildren become interested in mathematics, this book is dedicated to them: Megan Howden, Mia Davies, Lucie and Millie Vince, Ella and Henry Huber, and of course, my wife Heidi.

Preface

The first edition of *Vector Analysis for Computer Graphics* was published in 2007, and after submitting the manuscript to Springer, I came across geometric algebra. Since then, I have thought about it intensely, and written two books on the subject. Meanwhile, what I call traditional vector analysis, still exists, and continues to be taught throughout the world. So, I decided to write a 2nd edition, and here it is.

It has taken over a year to prepare the final files, because the originals had been created using MS Word, and I now use *LATEX*. This has meant retyping every equation and creating all the figures again!

The book's main objective is to introduce the subject of vector analysis to readers studying computer graphics, although it will be of interest to a wider readership. It includes most of the original topics, but I have extended various areas, and it still comprises eleven chapters. Chap. 1 covers the history of vector analysis, which answers most of the questions relating to the existence of the subject. I do believe that the stories behind the emergence of vectors are very relevant to the subject. Chap. 2 introduces dependent and independent equations, which are so important to vector analysis.

Chapter 3 covers the ideas behind vectors, building upon the concepts of linear equations, while Chap. 4 introduces the two products associated with vectors: the scalar and the vector product. The next two chapters are new: Chapter 5 introduces vector-valued functions and how they are differentiated and integrated; whilst Chap. 6 deals with vector differential operators: grad, div and curl. Chapter 7 develops the use of the grad differential operator by showing how tangent and normal vectors are computed for various curves and surfaces.

Chapters 8–10 cover applications to lines, planes and intersections. Chapter 11 examines three ways of rotating vectors in $\mathbb{R}^2$ and $\mathbb{R}^3$, finishing with quaternions.

I have thought very carefully about what to leave out, and what to include, and really hope that you are happy with the result.

Breinton, UK

John Vince

Contents

Chapter 1
History of Vector Analysis

1.1 Introduction

In this chapter I trace the emergence of the name *vector*, the invention of quaternions, and the branch of mathematics called *vector analysis*.

1.2 Vector

The word 'vector' derives from the Latin *vehĕre*, to carry, and was used in astronomical geometry during the seventeenth century. The *Lexicon Technicum: Or, An Universal Dictionary of Arts and Sciences* by the English writer, scientist, and Anglican priest John Harris [1666–1719], published in 1704, refers to 'vector':

> A Line supposed to be drawn from any Planet moving round a Center, or the Focus of an Ellipsis, to that Center or Focus, is by some Writers of the New Astronomy, called the Vector; because 'tis that Line by which the Planet seems to be carried round its Center [1].

'Vector' also appears in the form *radius vector*, which in French is *rayon vecteur*. It is found in *Traité de mécanique céleste* by the French scholar and polymath Pierre-Simon Laplace [1749–1827], and in *Application de l'Analyse à la Géométrie* by the French mathematician and inventor Gaspard Monge [1746–1818]:

> ... on nomme la droite *r* le rayon vecteur du point, et l'origine des coordonnées devient un pôle, d'où partent les rayons vecteurs des différens points de l'espace [2].

Although 'vector' was being used by various astronomers and mathematicians, no one had found a way of encoding a mathematical object that possessed magnitude and direction—this was eventually invented by the brilliant Irish mathematician, physicist and astronomer Sir William Rowan Hamilton [1805–1865].

© Springer-Verlag London Ltd., part of Springer Nature 2021
J. Vince, *Vector Analysis for Computer Graphics*,
https://doi.org/10.1007/978-1-4471-7505-6_1

1.3 Quaternion

Today, Hamilton is recognised as the inventor of quaternion algebra, which became the first non-commutative algebra to be discovered. One can imagine the elation he felt when finding a solution to a problem he had been thinking about for a decade!

The invention provided the first mathematical framework for manipulating vectorial quantities, although this was to be refined by the American theoretical physicist, chemist, and mathematician Josiah Willard Gibbs [1839–1903]. Although Hamilton had arrived at his invention through an algebraic route, it was obvious to him that quaternions had significant geometric potential, and he immediately started to explore their vectorial and rotational properties.

Unbeknown to Hamilton—and virtually everyone else at the time—the French social reformer, and brilliant recreational mathematician Benjamin Olinde Rodrigues [1795–1851], had already published a paper in 1840 describing how to represent two successive rotations about different axes, by a single rotation about a third axis [3]. What is more, Rodrigues expressed his solution using a scalar and a 3-D axis, which pre-empted Hamilton's own approach using a scalar and a vector, by three years! Simon Altmann has probably done more than any other person to set this record straight, and has published his views widely [4–7].

The very existence of complex numbers presented a tantalising question for mathematicians of the 18th and 19th centuries: could there be a 3-D equivalent? The answer to this question was not obvious, and many gifted mathematicians, including Gauss, Möbius, Grassmann, and Hamilton had been searching for the answer.

Hamilton's research is well documented and covers a period from the early 1830s to 1843, when he invented quaternions. And for a further 22 years, until his death in 1865, he was preoccupied with the subject. By 1833 he had shown that complex numbers form an algebra of couples, i.e. ordered pairs [8].

As a 2-D complex number is represented by $a + bi$, Hamilton conjectured that a 3-D complex number could be represented by the triple, $a + bi + cj$, where $i^2 = j^2 = -1$. However, the product of two such triples raises a problem with their algebraic expansion:

$$z_1 = a_1 + b_1 i + c_1 j$$
$$z_2 = a_2 + b_2 i + c_2 j$$
$$z_1 z_2 = (a_1 + b_1 i + c_1 j)(a_2 + b_2 i + c_2 j)$$
$$= a_1 a_2 + a_1 b_2 i + a_1 c_2 j$$
$$+ b_1 a_2 i + b_1 b_2 i^2 + b_1 c_2 i j + c_1 a_2 j + c_1 b_2 j i + c_1 c_2 j^2$$
$$= (a_1 a_2 - b_1 b_2 - c_1 c_2) + (a_1 b_2 + b_1 a_2)i + (a_1 c_2 + c_1 a_2)j$$
$$+ b_1 c_2 i j + c_1 b_2 j i.$$

The operation almost closes—apart from the terms involving ij and ji. Even if we assume that $ji = -ij$, we are still left with

$$(b_1 c_2 - c_1 b_2)ij.$$

This presented a real problem for Hamilton and he toiled for over a decade trying to resolve it. Then, on 16 October, 1843, whilst walking with his wife, Lady Hamilton, along the Royal Canal in Ireland to preside at a meeting of the Royal Irish Academy [9], a flash of inspiration came to him where he saw the solution as a quadruple, rather than a triple. Instead of using two imaginary terms, three terms provided the extra permutations necessary to resolve products like ij.

The solution was $z = a + bi + cj + dk$ where $i^2 = j^2 = k^2 = -1$. And because of the four terms, Hamilton gave the name *quaternion*. Hamilton took the opportunity to record the event in stone, by carving the rules into the wall of Broome bridge, which he was passing at the time. Although his original inscription has not withstood years of Irish weather, a more permanent plaque now replaces it.

When Hamilton invented quaternions, he also proposed all sorts of names such as *tensor*, *versor* and *vector* to describe their attributes. As the inventor, it was Hamilton's prerogative to choose whatever names he wanted, and at the time, such names were associated with the notation of the period. For example, he called the quaternion's real part a *scalar*, and the imaginary part a *vector*. However, today a vector does not have any imaginary associations, which has slightly confused how quaternions are interpreted.

1.4 Some History

Hamilton defined a quaternion q, and its associated rules as:

$$q = s + ia + jb + kc, \quad s, a, b, c \in \mathbb{R}$$

and

$$i^2 = j^2 = k^2 = ijk = -1,$$

$$ij = k, \quad jk = i, \quad ki = j,$$
$$ji = -k, \ kj = -i, \ ik = -j,$$

[10–12], but we tend to write quaternions as

$$q = s + ai + bj + ck.$$

Observe from Hamilton's rules how the occurrence of ij is replaced by k. The extra imaginary k term is key to the cyclic patterns $ij = k$, $jk = i$, and $ki = j$, which are very similar to the cross product of two unit Cartesian vectors:

$$\mathbf{i} \times \mathbf{j} = \mathbf{k}, \quad \mathbf{j} \times \mathbf{k} = \mathbf{i}, \quad \mathbf{k} \times \mathbf{i} = \mathbf{j}.$$

In fact, this similarity is no coincidence, as Hamilton also invented the scalar and vector products. However, although quaternions provided an algebraic framework to describe vectors, one must acknowledge that vectorial quantities had been studied for many years prior to Hamilton.

Hamilton also saw that the i, j, k terms could represent three Cartesian unit vectors $\mathbf{i}, \mathbf{j}$ and $\mathbf{k}$, which had to possess imaginary qualities. i.e. $\mathbf{i}^2 = -1$, etc., which didn't go down well with some mathematicians and scientists who were suspicious of the need to involve so many imaginary terms.

Hamilton's motivation to search for a 3-D equivalent of complex numbers was part algebraic, and part geometric. For if a complex number is represented by a couple and is capable of rotating points on the plane, then perhaps a *triple* rotates points in space. In the end, a triple had to be replaced by a a quadruple—a quaternion.

One can regard Hamilton's rules from two perspectives. The first, is that they are an algebraic consequence of combining three imaginary terms. The second, is that they reflect an underlying geometric structure of space. The latter interpretation was adopted by the Scottish mathematical physicist Peter Guthrie Tait [1831–1901], and outlined in his book *An Elementary Treatise on Quaternions* [13]. Tait's approach assumes three unit vectors $\mathbf{i}, \mathbf{j}, \mathbf{k}$ aligned with the x-, y-, z-axes respectively:

> The result of the multiplication of $\mathbf{i}$ into $\mathbf{j}$ or $\mathbf{ij}$ is defined to be the turning of $\mathbf{j}$ through a right angle in the plane perpendicular to $\mathbf{i}$ in the positive direction, in other words, the operation of $\mathbf{i}$ on $\mathbf{j}$ turns it round so as to make it coincide with $\mathbf{k}$; and therefore briefly $\mathbf{ij} = \mathbf{k}$.

> To be consistent it is requisite to admit that if $\mathbf{i}$ instead of operating on $\mathbf{j}$ had operated on any other unit vector perpendicular to $\mathbf{i}$ in the plane yz, it would have turned it through a right-angle in the same direction, so that $\mathbf{ik}$ can be nothing else than $-\mathbf{j}$.

> Extending to other unit vectors the definition which we have illustrated by referring to $\mathbf{i}$, it is evident that $\mathbf{j}$ operating on $\mathbf{k}$ must bring it round to $\mathbf{i}$, or $\mathbf{jk} = \mathbf{i}$.

Tait's explanation is illustrated in Fig. 1.1a–d. Figure 1.1a shows the original alignment of $\mathbf{i}, \mathbf{j}, \mathbf{k}$. Figure 1.1b shows the effect of turning $\mathbf{j}$ into $\mathbf{k}$. Figure 1.1c shows the turning of $\mathbf{k}$ into $\mathbf{i}$, and Fig. 1.1d shows the turning of $\mathbf{i}$ in to $\mathbf{j}$.

So far, there is no mention of imaginary quantities—we just have,

$$\mathbf{ij} = \mathbf{k}, \quad \mathbf{jk} = \mathbf{i}, \quad \mathbf{ki} = \mathbf{j},$$
$$\mathbf{ji} = -\mathbf{k}, \, \mathbf{kj} = -\mathbf{i}, \, \mathbf{ik} = -\mathbf{j}.$$

If we assume that these vectors obey the distributive and associative axioms of algebra, their imaginary qualities are exposed. For example:

$$\mathbf{ij} = \mathbf{k}$$

and multiplying throughout by $\mathbf{i}$:

$$\mathbf{iij} = \mathbf{ik} = -\mathbf{j}$$

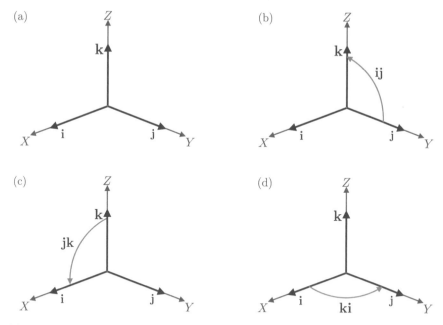

Fig. 1.1 Interpreting the products **ij**, **jk**, **ki**

therefore,

$$\mathbf{ii} = \mathbf{i}^2 = -1.$$

Similarly, we can show that $\mathbf{j}^2 = \mathbf{k}^2 = -1$.
    Next:

$$\mathbf{ijk} = \mathbf{i}(\mathbf{jk}) = \mathbf{ii} = \mathbf{i}^2 = -1.$$

Thus, simply by declaring the action of the cross-product, Hamilton's rules emerge, with all of their imaginary features. Tait also made the following observation:

> A very curious speculation, due to Servois, and published in 1813 in Gergonne's Annales is the only one, so far has been discovered, in which the slightest trace of an anticipation of Quaternions is contained. Endeavouring to extend to *space* the form $a + b\sqrt{-1}$ for the plane, he is guided by analogy to write a directed unit-line in space the form
>
> $$p \cos\alpha + q \cos\beta + r \cos\gamma,$$
>
> where α, β, γ are its inclinations to the three axes. He perceives easily that p, q, r must be *non-reals* : but, he asks, "seraient-elles imaginaires réductibles à la forme générale $A + B\sqrt{-1}$?" This could not be the answer. In fact they are the **i**, **j**, **k** of the Quaternion Calculus.

So the French mathematician François-Joseph Servois [1768–1847], was another person who came very close to discovering quaternions. Furthermore, both Tait and Hamilton were apparently unaware of the paper published by Rodrigues.

And it doesn't stop there. The brilliant German mathematician Carl Friedrich Gauss [1777–1855], was extremely cautious, and nervous of publishing anything too revolutionary, just in case he was ridiculed by fellow mathematicians. His diaries reveal that he had anticipated non-Euclidean geometry ahead of Nikolai Ivanovich Lobachevsky. And in a short note from his diary in 1819 [14], he reveals that he had identified a method of finding the product of two quadruples (a, b, c, d) and $(\alpha, \beta, \gamma, \delta)$ as,

$$(A, B, C, D) = (a, b, c, d)(\alpha, \beta, \gamma, \delta)$$
$$= (a\alpha - b\beta - c\gamma - d\delta, \ a\beta + b\alpha - c\delta + d\gamma$$
$$a\gamma + b\delta + c\alpha - d\beta, \ a\delta - b\gamma + c\beta + d\alpha).$$

At first glance, this result does not look like a quaternion product, but if we transpose the second and third coordinates of the quadruples, and treat them as quaternions, we have,

$$(A, B, C, D) = (a + ci + bj + dk)(\alpha + \gamma i + \beta j + \delta k)$$
$$= a\alpha - c\gamma - b\beta - d\delta + a(\gamma i + \beta j + \delta k)$$
$$+ \alpha(ci + bj + dk), \ (b\delta - d\beta)i + (d\gamma - c\delta)j + (c\beta - b\gamma)k,$$

which is identical to Hamilton's quaternion product! Furthermore, Gauss also realised that the product was non-commutative. However, he did not publish his findings, and it was left to Hamilton to invent quaternions for himself, publish his results and take the credit.

In 1881 and 1884, Josiah Willard Gibbs, at Yale University, printed his lecture notes on vector analysis for his students. Gibbs had cut the 'umbilical cord' between the real and vector parts of a quaternion and raised the 3-D vector as an independent object without any imaginary connotations. Gibbs also took on board the ideas of the German mathematician Hermann Günter Grassmann [1809–1877], who had been developing his own ideas for a vectorial system since 1832. Gibbs also defined the scalar and vector products using the relevant parts of the quaternion product. Finally, in 1901, a student of Gibbs, Edwin Bidwell Wilson, published Gibbs' notes in book form: *Vector Analysis*, which contains the notation in use today [15].

Quaternion algebra is definitely imaginary, yet simply by isolating the vector part and ignoring the imaginary rules, Gibbs was able to reveal a new branch of mathematics that exploded into vector analysis.

Hamilton and his supporters were unable to persuade their peers that quaternions could represent vectorial quantities, and eventually, Gibbs' notation won the day, and quaternions faded from the scene.

1.5 Summary

The invention of quaternions and vectors are important subjects, and I have attempted to summarise it within a few pages. However, if you wish to discover more about the subject, then I can recommend Michael Crowe's book: *A History of Vector Analysis* [16].

References

1. Allegranza M (2018) https://math.stackexchange.com/questions/2796910/origin-of-the-word-vector
2. Monge G (1807) Application de l'Analyse à la Géométrie, p24
3. Cheng H, Gupta KC (1989) An historical note on finite rotations. Trans ASME J Appl Mech 56(1):139–145
4. Altmann SL (1986) Rotations, quaternions and double groups. Dover Publications (2005). ISBN-13: 978-0-486-44518-2
5. Altmann SL (1989) Rodrigues, and the quaternion scandal. Mathe. Magz. 62(5):291–308
6. Altmann SL (1992) Icons and symmetries. Clarendon Press, Oxford
7. Altmann SL, Ortiz EL (eds) (2005) Mathematics and social utopias in France: Olinde Rodrigues and his Times, American mathematical society. Hist. Mathe 28. ISBN-10: 0-8218-3860-1, ISBN-13: 978-0-8218-3860-0
8. Hamilton WR (1833) The Mathematical papers of Sir William Rowan Hamilton. In: Conway AW, Synge JL (eds) Geometrical optics, vol I (1931); Conway AW McDonnell AJ (eds) Dynamics vol II (1940); Halberstam H, Ingram RE (eds) Algebra vol III (1940). Cambridge University Press, Cambridge
9. Hamilton WR, http://www-history.mcs.st-andrews.ac.uk/Mathematicians/Hamilton.html
10. Hamilton WR (1844) On quaternions: or a new system of imaginaries in algebra. Phil. Mag. 3rd ser. 25
11. Hamilton WR (1853) Lectures on quaternions. Hodges & Smith, Dublin
12. Hamilton WR (1899-1901) Elements of quaternions. In: Jolly CJ (ed) 2nd edn., vol 2. Longmans, Green & Co., London
13. Tait PG (1867) Elementary treatise on quaternions. Cambridge University Press, Cambridge
14. Gauss C F (1819) Mutation des Raumes, Carl Friedrich Gauss Werke, Achter Band, pp 357–361, König. Gesell. Wissen. Göttingen, 1900
15. Wilson EB (1901) Vector analysis. Yale University Press, London, UK
16. Crowe MJ (1994) A history of vector analysis. Dover Publications, New York

Chapter 2
Linear Equations

2.1 Introduction

In this chapter we review the ideas of linear equations as a foundation for the chapter on vector algebra. The explanations require a knowledge of matrix notation and the role of determinants, which the author has previously described in [1] and [2].

2.2 Systems of Linear Equations

In mathematics, an equation is said to be linear when it takes the form:

$$\lambda_1 x_1 + \lambda_2 x_2 + \cdots + \lambda_{n-1} x_{n-1} + \lambda_n x_n + c = 0.$$

Where the λ terms linearly scale the unknown, variable x terms, and c is an arbitrary constant. For example, (2.1), (2.2) and (2.3) are all linear equations:

$$2x + 3y - 2 = 0 \tag{2.1}$$
$$x - 2y + 5z = 0 \tag{2.2}$$
$$6x + 7y + 8z + 20t + 35 = 0. \tag{2.3}$$

Whereas, (2.4), (2.5) and (2.6), are all nonlinear equations:

$$2x^2 + 3y - 2 = 0 \tag{2.4}$$
$$x^3 - 2y + 5z^2 + 5 = 0 \tag{2.5}$$
$$6x^3 + 7y^2 + 8z + 20t^2 + 35 = 0. \tag{2.6}$$

© Springer-Verlag London Ltd., part of Springer Nature 2021
J. Vince, *Vector Analysis for Computer Graphics*,
https://doi.org/10.1007/978-1-4471-7505-6_2

A system of linear equations must comprise at least two or more linear equations, where one objective is to find a solution that satisfies **all** the equations. In order to find a unique solution, the equations must be *independent*, i.e. the equations must not be linearly related. Let's examine two systems of independent equations, and a system of dependent equations, for $\mathbb{R}^2$ and $\mathbb{R}^3$.

2.3 Independent Linear Equations in $\mathbb{R}^2$

A system of independent, linear equations in $\mathbb{R}^2$ is shown in (2.7), and graphed in Fig. 2.1:

$$\left. \begin{array}{l} x - 2y + 2 = 0 \\ x - y + 1 = 0. \end{array} \right\} \tag{2.7}$$

The graphs confirm that the two lines intersect at the point $x = 0$, $y = 1$, which is the only solution to (2.7).

The fact that two equations are independent, does not mean that a solution always exists. For instance, parallel lines never intersect, therefore their equations do not possess a solution. For example, (2.8) shows a system of independent, linear equations in $\mathbb{R}^2$, with identical slopes, but displaced horizontally, as shown in Fig. 2.2:

$$\left. \begin{array}{l} x - 2y + 2 = 0 \\ x - 2y + 4 = 0. \end{array} \right\} \tag{2.8}$$

Thus we must anticipate both possibilities when working with linear equations in $\mathbb{R}^2$.

Fig. 2.1 Graphs of
$x - 2y + 2 = 0$ (blue), and
$x - y + 1 = 0$ (green)

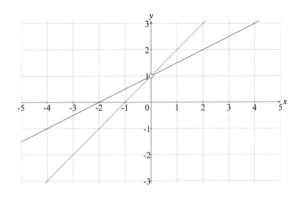

Fig. 2.2 Graphs of
$x - 2y + 2 = 0$ (blue), and
$x - 2y + 4 = 0$ (green)

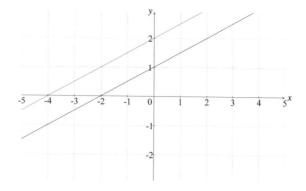

2.4   Dependent Linear Equations in $\mathbb{R}^2$

A system of dependent, linear equations in $\mathbb{R}^2$ is shown in (2.9), and graphed in Fig. 2.3, where the second equation is double the first:

$$\left. \begin{array}{l} x - 2y + 2 = 0 \\ 2x - 4y + 4 = 0. \end{array} \right\} \qquad (2.9)$$

As the two equations encode the same relationship between x and y, their graphs are identical and there is no single solution. For instance, $x = 2$, $y = 2$ satisfies both equations, so do $x = 0$, $y = 1$ and $x = -2$, $y = 0$. In fact, the two equations share an infinite number of solutions over their extent.

The three conditions: a single solution, no solution, and an infinite number of solutions, can be generalised to **any** system of linear equations.

Fig. 2.3 Graphs of
$x - 2y + 2 = 0$ (blue), and
$2x - 4y + 4 = 0$ (dashed
green)

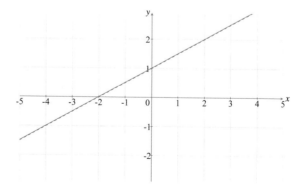

2.5 Consistent and Inconsistent Linear Equations

The terms *consistent* and *inconsistent* are often used in the context of linear equations. A consistent system of linear equations has at least one solution, i.e. intersecting or coincident lines, and an inconsistent system of linear equations has no solutions, i.e. parallel lines.

2.6 Independent Linear Equations in $\mathbb{R}^3$

A system of independent, linear equations in $\mathbb{R}^3$ is shown in (2.10):

$$\left. \begin{array}{r} y + z - 4 = 0 \\ z - 2 = 0 \\ x = 0 \end{array} \right\} \tag{2.10}$$

which represent three planes in 3-D space, as shown in Fig. 2.4. They intersect at a single point, whose coordinates are the common solution to the equations: $(0, 2, 2)$.

Once again, the fact that three equations are independent, does not mean that a solution always exists. For instance, parallel planes never intersect, therefore their equations do not possess a solution. For example, (2.11) shows a system of independent, linear equations in $\mathbb{R}^3$, two, of which, have identical slopes, but displaced in space:

$$\left. \begin{array}{r} y + z - 4 = 0 \\ y + z - 3 = 0 \\ x = 0. \end{array} \right\} \tag{2.11}$$

Figure 2.5 illustrates this combination of planes.

Fig. 2.4 Three planes intersecting at $(0, 2, 2)$

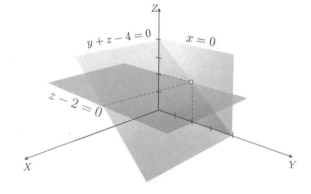

Fig. 2.5 Two parallel planes
and one other plane

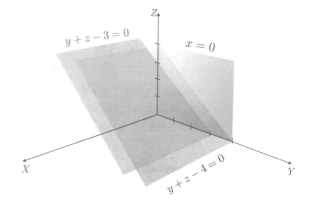

2.7 Dependent Linear Equations in $\mathbb{R}^3$

A system of dependent, linear equations in $\mathbb{R}^3$ is shown in (2.12), and graphed in
Fig. 2.6, which are all linearly related:

$$\left.\begin{array}{l} y + z - 4 = 0 \\ 2y + 2z - 8 = 0 \\ 3y + 3z - 12 = 0. \end{array}\right\} \tag{2.12}$$

As the three equations encode the same relationship between x, y and z, their
graphs are identical and there is no single solution. For instance, $x = 0$, $y = 2$, $z = 2$ satisfies all equations, so do $x = 1$, $y = 0$, $z = 4$ and $x = 2$, $y = 4$, $z = 0$. In
fact, the three equations share an infinite number of solutions over their extent.

Fig. 2.6 Three coincident
planes

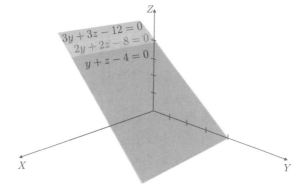

2.8   Matrix Notation

2.8.1   Linear Equations in $\mathbb{R}^2$

Matrix notation provides a powerful way of representing linear equations, and a simple way of solving them. For example, the linear equations (2.13) are represented in matrix form as shown in (2.14):

$$\left. \begin{array}{c} a_1 x + a_2 y = c_1 \\ b_1 x + b_2 y = c_2 \end{array} \right\} \tag{2.13}$$

$$\begin{bmatrix} a_1 & a_2 \\ b_1 & b_2 \end{bmatrix} \begin{bmatrix} x \\ y \end{bmatrix} = \begin{bmatrix} c_1 \\ c_2 \end{bmatrix}. \tag{2.14}$$

The 2×2 matrix in (2.14) encodes two vital bits of information about the original equations: whether they are independent, or not, and the inverse of the matrix, if one exists. Let us start with the independence of the equations.

2.8.2   Second-Order Determinant

Consider the second-order matrix of (2.14) whose determinant is written:

$$\begin{vmatrix} a_1 & a_2 \\ b_1 & b_2 \end{vmatrix} = a_1 b_2 - a_2 b_1$$

which is a scaling factor when the matrix is used as a geometric transform. This is shown by applying (2.14) to the four points of a unit square, we have:

$$\begin{bmatrix} a_1 & a_2 \\ b_1 & b_2 \end{bmatrix} \begin{bmatrix} 0 \\ 0 \end{bmatrix} = \begin{bmatrix} 0 \\ 0 \end{bmatrix}$$

$$\begin{bmatrix} a_1 & a_2 \\ b_1 & b_2 \end{bmatrix} \begin{bmatrix} 1 \\ 0 \end{bmatrix} = \begin{bmatrix} a_1 \\ b_1 \end{bmatrix}$$

$$\begin{bmatrix} a_1 & a_2 \\ b_1 & b_2 \end{bmatrix} \begin{bmatrix} 1 \\ 1 \end{bmatrix} = \begin{bmatrix} a_1 + a_2 \\ b_1 + b_2 \end{bmatrix}$$

$$\begin{bmatrix} a_1 & a_2 \\ b_1 & b_2 \end{bmatrix} \begin{bmatrix} 0 \\ 1 \end{bmatrix} = \begin{bmatrix} a_2 \\ b_2 \end{bmatrix}.$$

Fig. 2.7 The area marked A
is the scaled unit square

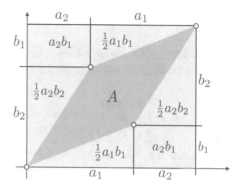

The matrix transforms the original unit square to the parallelogram shown in Fig. 2.7, whose area is calculated as follows:

$$
\begin{aligned}
A &= (a_1 + a_2)(b_1 + b_2) - \tfrac{1}{2}a_1b_1 - a_2b_1 - \tfrac{1}{2}a_2b_2 - \tfrac{1}{2}a_1b_1 - a_2b_1 - \tfrac{1}{2}a_2b_2 \\
&= (a_1 + a_2)(b_1 + b_2) - a_1b_1 - a_2b_1 - a_2b_2 - a_2b_1 \\
&= a_1b_1 + a_1b_2 + a_2b_1 + a_2b_2 - a_1b_1 - a_2b_1 - a_2b_2 - a_2b_1 \\
&= a_1b_2 - a_2b_1.
\end{aligned}
$$

This shows that the matrix representing two independent linear equations, scales the area of a shape by the absolute value of its determinant.

If the equations are linearly related, as shown in (2.15):

$$
\left.
\begin{aligned}
a_1x + a_2y &= c_1 \\
\lambda a_1x + \lambda a_2y &= \lambda c_1
\end{aligned}
\right\}
\tag{2.15}
$$

where λ is some scaling factor, then the determinant of the associated matrix collapses to zero, as shown in (2.16):

$$
\begin{bmatrix} a_1 & a_2 \\ \lambda a_1 & \lambda a_2 \end{bmatrix}
\begin{bmatrix} x \\ y \end{bmatrix} =
\begin{bmatrix} c_1 \\ \lambda c_1 \end{bmatrix}
$$

$$
\begin{vmatrix} a_1 & a_2 \\ \lambda a_1 & \lambda a_2 \end{vmatrix} = 0.
\tag{2.16}
$$

Thus a determinant can be used to detect linear dependency.

2.8.3   Third-Order Determinant

The French mathematician Pierre Sarrus [1798–1861] discovered a simple way to compute the value of a third-order determinant, as shown in (2.17):

$$\mathbf{A} = \begin{bmatrix} a_1 & a_2 & a_3 \\ b_1 & b_2 & b_3 \\ c_1 & c_2 & c_3 \end{bmatrix}$$

$$\det \mathbf{A} = a_1 b_2 c_3 + a_2 b_3 c_1 + a_3 b_1 c_2 - a_1 b_3 c_2 - a_2 b_1 c_3 - a_3 b_2 c_1. \qquad (2.17)$$

For example, given

$$\mathbf{A} = \begin{bmatrix} 2 & 3 & 5 \\ 7 & 11 & 13 \\ 17 & 19 & 23 \end{bmatrix}$$

then,

$$\begin{aligned} \det \mathbf{A} &= (2 \times 11 \times 23) + (3 \times 13 \times 17) + (5 \times 7 \times 19) \\ &\quad - (2 \times 13 \times 19) - (3 \times 7 \times 23) - (5 \times 11 \times 17) \\ &= 506 + 663 + 665 - 494 - 483 - 935 \\ &= -78. \end{aligned}$$

If the equations are linearly related, as shown in (2.18):

$$\left. \begin{aligned} a_1 x + a_2 y + a_3 z &= c_1 \\ b_1 x + b_2 y + b_3 z &= c_2 \\ \lambda a_1 x + \lambda a_2 y + \lambda a_3 z &= \lambda c_1 \end{aligned} \right\} \qquad (2.18)$$

where λ is some scaling factor, then the determinant of the associated matrix collapses to zero, as shown in (2.19):

$$\mathbf{A} = \begin{bmatrix} a_1 & a_2 & a_3 \\ b_1 & b_2 & b_3 \\ \lambda a_1 & \lambda a_2 & \lambda a_3 \end{bmatrix}$$

$$\det \mathbf{A} = \lambda a_1 a_3 b_2 + \lambda a_1 a_2 b_3 + \lambda a_2 a_3 b_1 - \lambda a_1 a_2 b_3 - \lambda a_2 a_3 b_1 - \lambda a_1 a_3 b_2 = 0. \qquad (2.19)$$

2.9   Solving Linear Equations

In order to demonstrate how linear equations can be solved, we require the use of two matrices: the identity matrix and the inverse matrix. The identity matrix is shown in (2.20) for a 2 × 2 matrix:

$$\begin{bmatrix} 1 & 0 \\ 0 & 1 \end{bmatrix} \tag{2.20}$$

and in (2.21) for a 3 × 3 matrix:

$$\begin{bmatrix} 1 & 0 & 0 \\ 0 & 1 & 0 \\ 0 & 0 & 1 \end{bmatrix} . \tag{2.21}$$

The identity matrix is the matrix equivalent of the scalar 1, as shown in (2.22) and (2.23):

$$\begin{bmatrix} 1 & 0 \\ 0 & 1 \end{bmatrix} \begin{bmatrix} x \\ y \end{bmatrix} = \begin{bmatrix} x \\ y \end{bmatrix} \tag{2.22}$$

$$\begin{bmatrix} 1 & 0 & 0 \\ 0 & 1 & 0 \\ 0 & 0 & 1 \end{bmatrix} \begin{bmatrix} x \\ y \\ z \end{bmatrix} = \begin{bmatrix} x \\ y \\ z \end{bmatrix} . \tag{2.23}$$

In scalar arithmetic we can divide two quantities x and y, either by writing x/y or $x \cdot y^{-1}$. In matrix notation, the only way to divide matrix $\mathbf{A}$ by matrix $\mathbf{B}$, is to multiply $\mathbf{A}$ by $\mathbf{B}^{-1}$, that is, when it is possible to compute the inverse.

Similarly, as the product $y \cdot y^{-1} = 1$, the product of the two matrices $\mathbf{B} \cdot \mathbf{B}^{-1}$, equals the identity matrix.

For a 2 × 2 matrix $\mathbf{A}$ the inverse is $\mathbf{A}^{-1}$ (2.24):

$$\mathbf{A} = \begin{bmatrix} a & b \\ c & d \end{bmatrix} , \quad \text{then} \quad \mathbf{A}^{-1} = \frac{1}{\det \mathbf{A}} \begin{bmatrix} d & -b \\ -c & a \end{bmatrix} . \tag{2.24}$$

For a 3 × 3 matrix $\mathbf{A}$ the inverse is $\mathbf{A}^{-1}$ (2.25):

$$\mathbf{A} = \begin{bmatrix} a & b & c \\ d & e & f \\ g & h & j \end{bmatrix}$$

$$\mathbf{A}^{-1} = \frac{1}{\det \mathbf{A}} \begin{bmatrix} (ej - fh) & -(bj - ch) & (bf - ce) \\ -(dj - gf) & (aj - gc) & -(af - dc) \\ (dh - ge) & -(ah - gb) & (ae - bd) \end{bmatrix} . \tag{2.25}$$

2.9.1   Solving Linear Equations in $\mathbb{R}^2$

Knowing that there are three possible solutions to a system of linear equations permits us to design three strategies to solve such equations. Sometimes it is obvious from the equations that they are dependent, as in the case of (2.9), where the second equation is double the first. However, it is not obvious that (2.26) is a dependent system:

$$\left.\begin{array}{r}123x - 45y + 937 = 0 \\ 2091x - 765y + 15,929 = 0.\end{array}\right\} \qquad (2.26)$$

The second equation is 17 times the first, and a determinant will always detect this condition:

$$\begin{vmatrix} 123 & -45 \\ 2091 & -765 \end{vmatrix} = -123 \times 765 - (-45 \times 2091) = -94095 + 94095 = 0.$$

Using (2.7) and moving the constants across, we have:

$$x - 2y = -2$$
$$x - y = -1$$
$$\begin{bmatrix} 1 & -2 \\ 1 & -1 \end{bmatrix}\begin{bmatrix} x \\ y \end{bmatrix} = \begin{bmatrix} -2 \\ -1 \end{bmatrix}. \qquad (2.27)$$

Using (2.24) we have:

$$\mathbf{A} = \begin{bmatrix} 1 & -2 \\ 1 & -1 \end{bmatrix}$$
$$\mathbf{A}^{-1} = \frac{1}{1}\begin{bmatrix} -1 & 2 \\ -1 & 1 \end{bmatrix}. \qquad (2.28)$$

Multiplying both sides of (2.27) by (2.28), we have:

$$\begin{bmatrix} -1 & 2 \\ -1 & 1 \end{bmatrix}\begin{bmatrix} 1 & -2 \\ 1 & -1 \end{bmatrix}\begin{bmatrix} x \\ y \end{bmatrix} = \begin{bmatrix} -1 & 2 \\ -1 & 1 \end{bmatrix}\begin{bmatrix} -2 \\ -1 \end{bmatrix}$$
$$\begin{bmatrix} 1 & 0 \\ 0 & 1 \end{bmatrix}\begin{bmatrix} x \\ y \end{bmatrix} = \begin{bmatrix} -1 & 2 \\ -1 & 1 \end{bmatrix}\begin{bmatrix} -2 \\ -1 \end{bmatrix}$$
$$\begin{bmatrix} x \\ y \end{bmatrix} = \begin{bmatrix} 0 \\ 1 \end{bmatrix}.$$

Which confirms that the solution to (2.7) is $x = 0$, $y = 1$.

Independent equations do not imply that the determinant of the associated matrix is always non-zero; in the case of (2.8), representing parallel lines, the determinant is zero:

$$x - 2y + 2 = 0$$
$$x - 2y + 4 = 0$$
$$\begin{bmatrix} 1 & -2 \\ 1 & -2 \end{bmatrix} \begin{bmatrix} x \\ y \end{bmatrix} = \begin{bmatrix} -2 \\ -4 \end{bmatrix}$$
$$\begin{vmatrix} 1 & -2 \\ 1 & -2 \end{vmatrix} = 0.$$

2.9.2 Solving Linear Equations in $\mathbb{R}^3$

Let's solve (2.10) using (2.25), even though the solution is clear, as we know that $x = 0$, and $z = 2$, it follows that $y = 2$. We use (2.25) to calculate the inverse matrix $\mathbf{A}^{-1}$:

$$y + z - 4 = 0$$
$$z - 2 = 0$$
$$x = 0$$
$$\begin{bmatrix} 0 & 1 & 1 \\ 0 & 0 & 1 \\ 1 & 0 & 0 \end{bmatrix} \begin{bmatrix} x \\ y \\ z \end{bmatrix} = \begin{bmatrix} 4 \\ 2 \\ 0 \end{bmatrix} \qquad (2.29)$$
$$\mathbf{A} = \begin{bmatrix} 0 & 1 & 1 \\ 0 & 0 & 1 \\ 1 & 0 & 0 \end{bmatrix}$$
$$\mathbf{A}^{-1} = \begin{bmatrix} 0 & 0 & 1 \\ 1 & -1 & 0 \\ 0 & 1 & 0 \end{bmatrix}. \qquad (2.30)$$

Multiplying both sides of (2.29) by (2.30), we have:

$$
\begin{bmatrix} 0 & 0 & 1 \\ 1 & -1 & 0 \\ 0 & 1 & 0 \end{bmatrix}
\begin{bmatrix} 0 & 1 & 1 \\ 0 & 0 & 1 \\ 1 & 0 & 0 \end{bmatrix}
\begin{bmatrix} x \\ y \\ z \end{bmatrix}
=
\begin{bmatrix} 0 & 0 & 1 \\ 1 & -1 & 0 \\ 0 & 1 & 0 \end{bmatrix}
\begin{bmatrix} 4 \\ 2 \\ 0 \end{bmatrix}
$$

$$
\begin{bmatrix} 1 & 0 & 0 \\ 0 & 1 & 0 \\ 0 & 0 & 1 \end{bmatrix}
\begin{bmatrix} x \\ y \\ z \end{bmatrix}
=
\begin{bmatrix} 0 & 0 & 1 \\ 1 & -1 & 0 \\ 0 & 1 & 0 \end{bmatrix}
\begin{bmatrix} 4 \\ 2 \\ 0 \end{bmatrix}
$$

$$
\begin{bmatrix} x \\ y \\ z \end{bmatrix}
=
\begin{bmatrix} 0 & 0 & 1 \\ 1 & -1 & 0 \\ 0 & 1 & 0 \end{bmatrix}
\begin{bmatrix} 4 \\ 2 \\ 0 \end{bmatrix}
$$

$$
\begin{bmatrix} x \\ y \\ z \end{bmatrix}
=
\begin{bmatrix} 0 \\ 2 \\ 2 \end{bmatrix}. \tag{2.31}
$$

Equation (2.31) confirms that the solution to (2.10) is $x = 0, \ y = 2, \ z = 2$.

2.10   Summary

Hopefully, this chapter has clarified some important ideas about linear equations, in particular, independent and dependent linear equations. One can see that matrices and determinants are very useful in representing linear equations, and will be employed in the chapters on problem solving.

References

1. Vince J (2020) Foundation mathematics for computer science, 2th edn. Springer. ISBN 978-3-030-42077-2
2. Vince J (2017) Mathematics for computer graphics, 5th edn. Springer. ISBN 978-1-4471-7334-2

Chapter 3
Vector Algebra

3.1 Introduction

This chapter has six sections, and begins by reviewing the mathematical objects of groups, rings and fields, as these are employed in the description of vectors. The second section is about space, and covers Euclidean space, vector space and normed vector space. The third section is concerned with vector definition, vector notation, addition and various vector features. The fourth section describes linearly dependent and independent vectors and shows the role of the determinant in distinguishing between the two conditions. The fifth section covers the span of a set of vectors, and section six introduces the ideas of basis vectors and dimension.

3.2 Groups, Rings and Fields

3.2.1 Groups

When certain arithmetic operations are applied to members of a set we can secure closure, non-closure, or the result is undefined. For example, we secure closure for most arithmetic operations with the set $\mathbb{R}$, but division by zero causes problems, and is undefined.

When combining sets with arithmetic operations, it is convenient to create another entity: a *group*, which is a set, together with the axioms describing how elements of the set are combined. The set might contain numbers, matrices, vectors, quaternions, polynomials, etc., and are represented below as a, b and c.

The axioms employ the 'o' symbol to represent any binary operation such as $+, -, \times$. And a group is formed from a set and a binary operation. For example, we may wish to form a group of integers under addition: $(\mathbb{Z}, +)$, or we may wish to examine whether quaternions form a group under the operation of multiplication: $(\mathbb{H}, \times)$.

© Springer-Verlag London Ltd., part of Springer Nature 2021
J. Vince, *Vector Analysis for Computer Graphics*,
https://doi.org/10.1007/978-1-4471-7505-6_3

To be a group, **all** the following axioms **must** hold for the set S. In particular, there must be a special *identity element* $e \in S$, and for each $a \in S$ there must exist an *inverse element* $a^{-1} \in S$, so that the following axioms are satisfied:

$$\begin{aligned}
\text{Closure:} \quad & a \circ b \in S, & a, b \in S. \\
\text{Associativity:} \quad & (a \circ b) \circ c = a \circ (b \circ c), & a, b, c \in S. \\
\text{Identity element:} \quad & a \circ e = e \circ a = a, & a, e \in S. \\
\text{Inverse element:} \quad & a \circ a^{-1} = a^{-1} \circ a = e, & a, a^{-1}, e \in S.
\end{aligned}$$

We describe a group as $(S, \circ)$, where S is the set and '$\circ$' the operation. For instance, $(\mathbb{Z}, +)$ is the group of integers under the operation of addition, and $(\mathbb{R}, \times)$ is the group of reals under the operation of multiplication.

Let's bring these axioms to life with three examples.

$(\mathbb{Z}, +)$: The integers $\mathbb{Z}$ form a group under the operation of addition:

$$\begin{aligned}
\text{Closure:} \quad & -23 + 24 = 1. \\
\text{Associativity:} \quad & (2 + 3) + 4 = 2 + (3 + 4) = 9. \\
\text{Identity:} \quad & 2 + 0 = 0 + 2 = 2. \\
\text{Inverse:} \quad & 2 + (-2) = (-2) + 2 = 0.
\end{aligned}$$

$(\mathbb{Z}, \times)$: The integers $\mathbb{Z}$ do **not** form a group under multiplication:

$$\begin{aligned}
\text{Closure:} \quad & -2 \times 4 = -8. \\
\text{Associativity:} \quad & (2 \times 3) \times 4 = 2 \times (3 \times 4) = 24. \\
\text{Identity:} \quad & 2 \times 1 = 1 \times 2 = 2. \\
\text{Inverse:} \quad & 2^{-1} = 0.5 \quad (0.5 \notin \mathbb{Z}).
\end{aligned}$$

Also, the integer 0 has no inverse.

$(\mathbb{Q}, \times)$: The group of non-zero rational numbers form a group under multiplication:

$$\begin{aligned}
\text{Closure:} \quad & \tfrac{2}{5} \times \tfrac{2}{3} = \tfrac{4}{15}. \\
\text{Associativity:} \quad & \left(\tfrac{2}{5} \times \tfrac{2}{3}\right) \times \tfrac{1}{2} = \tfrac{2}{5} \times \left(\tfrac{2}{3} \times \tfrac{1}{2}\right) = \tfrac{2}{15}. \\
\text{Identity:} \quad & \tfrac{2}{3} \times \tfrac{1}{1} = \tfrac{1}{1} \times \tfrac{2}{3} = \tfrac{2}{3}. \\
\text{Inverse:} \quad & \tfrac{2}{3} \times \tfrac{3}{2} = \tfrac{1}{1} \quad \left(\text{where } \tfrac{3}{2} = \left(\tfrac{2}{3}\right)^{-1}\right).
\end{aligned}$$

3.2.2 Abelian Group

An *abelian group*, named after the Norwegian mathematician Neils Henrik Abel [1802–1829], is a group where the order of elements does not influence the result. i.e. the group is commutative. Thus there are five axioms: closure, associativity, identity element, inverse element, and commutativity:

$$\text{Commutativity: } a \circ b = b \circ a, \quad a, b \in S.$$

For example, the set of integers forms an abelian group under ordinary addition $(\mathbb{Z}, +)$.

3.2.3 Rings

A *ring* is an extended group, where we have a set of objects which can be added and multiplied together, subject to some precise axioms. There are rings of real numbers, complex numbers, integers, matrices, equations, polynomials, etc. A ring is formally defined as a system where $(S, +)$ and $(S, \times)$ are abelian groups and the distributive axioms:

$$\text{Additive associativity: } a + (b + c) = (a + b) + c, \qquad a, b, c \in S.$$
$$\text{Multiplicative associativity: } a \times (b \times c) = (a \times b) \times c, \qquad a, b, c \in S.$$
$$\text{Distributivity: } a \times (b + c) = (a \times b) + (a \times c) \text{ and}$$
$$(a + b) \times c = (a \times c) + (b \times c), \qquad a, b, c \in S.$$

For example, we already know that the integers $\mathbb{Z}$ form a group under the operation of addition, but they also form a ring, as the set satisfies the above axioms:

$$2 \times (3 \times 4) = (2 \times 3) \times 4$$
$$2 \times (3 + 4) = (2 \times 3) + (2 \times 4)$$
$$(2 + 3) \times 4 = (2 \times 4) + (3 \times 4).$$

3.2.4 Fields

Although rings support addition and multiplication, they do not necessarily support division. However, as division is such an important arithmetic operation, the *field* was created to support it, with one proviso: division by zero is not permitted. Thus we have fields of real numbers $\mathbb{R}$, rational numbers $\mathbb{Q}$, and the complex numbers $\mathbb{C}$.

It follows that every field is a ring, but not every ring is a field.

3.2.5   Division Ring

A *division ring* or *division algebra*, is a ring in which every element has an inverse element, with the proviso that the element is non-zero. The algebra also supports non-commutative multiplication. Here is a formal description of the division ring $(S, +, \times)$:

$$
\begin{array}{rll}
\text{Additive associativity:} & (a + b) + c = a + (b + c), & a, b, c \in S. \\
\text{Additive commutativity:} & a + b = b + a, & a, b \in S. \\
\text{Additive identity 0:} & 0 + a = a, & a, 0 \in S. \\
\text{Additive inverse:} & a + (-a) = 0, & a, -a \in S. \\
\text{Multiplicative associativity:} & (a \times b) \times c = a \times (b \times c), & a, b, c \in S. \\
\text{Multiplicative identity 1:} & 1 \times a = a, & a, 1 \in S. \\
\text{Multiplicative inverse:} & a \times a^{-1} = 1, & a, a^{-1} \in S,\ a \neq 0. \\
\text{Distributivity:} & a \times (b + c) = (a \times b) + (a \times c) \text{ and} & \\
& (b + c) \times a = (b \times a) + (c \times a), & a, b, c \in S.
\end{array}
$$

The German mathematician Adolf Hurwitz [1859–1919], proved that there are only three associative division algebras: real numbers $\mathbb{R}$, complex numbers $\mathbb{C}$, and quaternions $\mathbb{H}$. This proof was published posthumously in 1923.

3.3   Space

3.3.1   Euclidean Space

Since the days of Euclid [c. 325 BC], 3-D *Euclidean space* has become an abstraction of the space we inhabit, and 2-D Euclidean space is a special case restricted to a flat two-dimensional plane. Both of these spaces are linear and are used to describe real-world phenomena from atoms to galaxies. However, current thinking is forcing us to consider other spatial models where at the smallest scales space is granular [1], and at the largest scales spacetime is curved [2].

Historically, a Euclidean space is formed by starting with a set of axioms which are used to construct other observations. Euclid's postulates are [3]:

1. A straight line segment may be drawn from any given point to any other.
2. A straight line may be extended to any finite length.
3. A circle may be described with any given point as its centre and any distance as its radius.
4. All right angles are congruent.
5. If a straight line intersects two other straight lines, and so makes the two interior angles on one side of it together less than two right angles, then the other straight lines will meet at a point if extended far enough on the side on which the angles are less than two right angles.

For example, these postulates permit one to show that a triangle's interior angles sum to two right angles.

Although modern mathematics still acknowledges the existence of the above postulates, Euclidean space has a more abstract definition where it is associated with the set of real numbers $\mathbb{R}$. Generally, an n-dimensional Euclidean space is equivalent to the set $\mathbb{R}^n$, where each point is equivalent to an n-tuple of $\mathbb{R}$. Thus in two-dimensional space, two real numbers uniquely define a single point, and in three-dimensional space, three real numbers are required. This definition permits the existence of Euclidean spaces with higher dimensions than three, which do not concern us in this book.

As the set of real numbers $\mathbb{R}$ lends itself to the actions of addition, subtraction and multiplication, so must Euclidean space, which means that Euclidean vectors can be added, subtracted and multiplied together using the scalar (dot) product.

3.3.2   Vector Space

An algebra is a mathematical structure comprising a set, and the operations of multiplication, addition and scalar multiplication by elements of the associated field. For example, algebra over the field of real numbers has well-defined operations of addition, multiplication and division, expressed in the axioms of associativity, commutativity and distributivity. An algebra over a field is also a *vector space* that possesses a bilinear product, and we often use the expression: "Let V be a vector space over a field F."

A bilinear form is defined as follows:

Definition: Let V be a vector space over a field F. A bilinear form B on V is a function of two elements r and s of V, and an element f of F, satisfying the following axioms:

$$B(r_1 + r_2, \ s) = B(r_1, \ s) + B(r_2, \ s)$$
$$B(r, \ s_1 + s_2) = B(r, \ s_1) + B(r, \ s_2)$$
$$B(fr, \ s) = f B(r, \ s)$$
$$B(r, \ fs) = f B(r, \ s).$$

The bilinear product is also a feature of the algebra of real numbers:

$$\text{Distributivity:} \quad a(b+c) = (ab) + (ac) \text{ and}$$
$$(b+c)a = (ba) + (ca), \qquad a, b, c \in \mathbb{R}.$$

The real numbers $\mathbb{R}$, can be seen as a one-dimensional vector space over the field $\mathbb{R}$, and thus are a one-dimensional algebra over itself. The complex numbers $\mathbb{C}$, can be seen as a two-dimensional vector space over the field $\mathbb{R}$, and thus are a two-dimensional algebra over $\mathbb{R}$. Although one- and two-dimensional vector spaces support division, there is no three-dimensional division algebra. However, quaternions form an algebra over the real numbers and support quaternion division. Thus the quaternions $\mathbb{H}$, can be seen as a four-dimensional vector space over the field $\mathbb{R}$.

3.3.3 Normed Vector Space

A *normed vector space* is a vector space on which a norm is defined. For example, the Euclidean norm is a real-valued function that determines the absolute magnitude of an element:

$$\text{real number } x : \ |x| = \sqrt{x^2}$$
$$\text{complex number } z = a + bi : \ |z| = \sqrt{a^2 + b^2}$$
$$\text{quaternion } q = [s, \ \lambda\hat{\mathbf{v}}] : \ |q| = \sqrt{s^2 + \lambda^2}$$
$$\text{Euclidean vector } \mathbf{a} = [x \ \ y \ \ z] : \ ||\mathbf{a}|| = \sqrt{x^2 + y^2 + z^2}.$$

3.4 Vector Definition

3.4.1 Geometric Vector

A *geometric vector* encodes a numerical quantity that has magnitude and direction, which enables it to represent a force, velocity, acceleration, etc. The most popular form is the Euclidean vector, which is used to solve problems in two and three dimensions, and takes the form of an ordered list (*n*-tuple) of scalars.

3.4.2 Euclidean Vector

A *Euclidean vector* is a geometric vector used to solve geometric problems in two and three dimensions.

3.4.3 Position Vector

A *position vector* is generally used to locate a point in space, and can be visualised with its tail at the origin, and its head at the desired point. It provides a simple strategy for solving complex 2-D and 3-D geometric problems.

3.4.4 n-tuple

Vectors are defined in different ways from a directed line to an object belonging to a multi-dimensional abstract space. A definition that permits us to explore directed lines or abstract spaces is that of ordered *n-tuples*, where a tuple is an ordered list of elements. The number of elements n, determines the dimension of a vector, thus a 2-tuple belongs to $\mathbb{R}^2$ (2-D space), a 3-tuple belongs to $\mathbb{R}^3$ (3-D space) and an n-tuple belongs to $\mathbb{R}^n$ (nD space). The following are all vectors using using the idea of tuples:

$$\mathbf{a} = [1 \quad 2] \in \mathbb{R}^2, \quad \mathbf{b} = [23 \quad 5 \quad 41] \in \mathbb{R}^3, \quad \mathbf{c} = [27 \quad 03 \quad 20 \quad 11] \in \mathbb{R}^4.$$

It is up to us what interpretation we assign to these elements: we can visualise the n-tuple as a position vector locating a point relative to the origin; the elements of the n-tuple can be Cartesian offsets of a floating directed line, or a list of numbers associated with an imaginary space.

3.4.5 Vector Notation

Vectors are notated in various ways:
- Boldface $\mathbf{a}$, $\mathbf{b}$, $\mathbf{c}$.
- Arrows $\overrightarrow{AB}$, $\overrightarrow{BC}$, $\overrightarrow{AC}$.

Authors may employ different bracket styles to enclose a vector's elements:
- Square [1 2 3].
- Parentheses (1, 2, 3).
- Angles < 1, 2, 3 >.

Finally, a vector's elements may or may not be separated with commas. A personal preference is as follows:

$$\mathbf{b} = \begin{bmatrix} 1 \\ 2 \\ 3 \end{bmatrix}, \quad \text{or} \quad \mathbf{a} = [4 \quad 5 \quad 6].$$

and used throughout this book.

3.4.6   *Column and Row Vectors and the Transpose Operation*

Vectors are expressed either as a *column* or *row* of elements:

$$\mathbf{b} = \begin{bmatrix} 1 \\ 2 \\ 3 \end{bmatrix}, \quad \text{or} \quad \mathbf{a} = [4 \ \ 5 \ \ 6]$$

and are moved between the two formats using the *transpose* operation:

$$\mathbf{b}^{\mathrm{T}} = [1 \ \ 2 \ \ 3], \quad \text{or} \quad \mathbf{a}^{\mathrm{T}} = \begin{bmatrix} 4 \\ 5 \\ 6 \end{bmatrix}.$$

One of the uses of the transpose operation is the product between two column vectors. For example, the product **bc** is not permitted:

$$\text{where} \quad \mathbf{b} = \begin{bmatrix} 1 \\ 2 \\ 3 \end{bmatrix}, \quad \mathbf{c} = \begin{bmatrix} 5 \\ 6 \\ 7 \end{bmatrix}$$

whereas, using the transpose operation the product becomes legal:

$$\mathbf{b}^{\mathrm{T}}\mathbf{c} = [1 \ \ 2 \ \ 3] \begin{bmatrix} 5 \\ 6 \\ 7 \end{bmatrix}.$$

We'll say more about vector products later on.

Column vectors arise in matrix algebra, especially in the context of systems of linear equations. For example, the set of Equ. (3.1) is represented in matrix form with column vectors (3.2):

$$\left. \begin{aligned} 3x + 4y + 5z &= 12 \\ 2x + 3y - 6z &= 10 \\ 1x - 7y + 3z &= 8 \end{aligned} \right\} \tag{3.1}$$

$$\begin{bmatrix} 3 & 4 & 5 \\ 2 & 3 & -6 \\ 1 & -7 & 3 \end{bmatrix} \begin{bmatrix} x \\ y \\ z \end{bmatrix} = \begin{bmatrix} 12 \\ 10 \\ 8 \end{bmatrix}. \tag{3.2}$$

3.4.7 Vector Magnitude

The *magnitude* or *modulus* of a vector is given by the associated vector space norm, which for Euclidean vectors is its length. For example:

$$\mathbf{a} = [3 \quad 4] \quad \text{and} \quad \|\mathbf{a}\| = \sqrt{3^2 + 4^2} = 5$$
$$\mathbf{b} = [4 \quad 5 \quad 6] \quad \text{and} \quad \|\mathbf{b}\| = \sqrt{4^2 + 5^2 + 6^2} = \sqrt{77}.$$

3.4.8 Vector Addition

Vector addition is just a question of adding the corresponding elements of the vectors, which must be of the same dimension. Given two vectors of dimension n:

$$\mathbf{a} = [a_1 \quad a_2 \quad \cdots \quad a_n]$$
$$\mathbf{b} = [b_1 \quad b_2 \quad \cdots \quad b_n]$$
$$\mathbf{a} + \mathbf{b} = [a_1 + b_1 \quad a_2 + b_2 \quad \cdots \quad a_n + b_n].$$

For example:
$$\mathbf{a} = [8 \quad 2], \quad \mathbf{b} = [3 \quad 3], \quad \mathbf{a} + \mathbf{b} = [11 \quad 5].$$

See Fig. 3.1.

3.4.9 Unit Vector

A *unit vector* has a magnitude of unity and is readily scaled to any required length by multiplying it by a scalar. A unit vector is given a 'hat' symbol $\hat{\mathbf{a}}$, to distinguish

Fig. 3.1 The addition of two vectors

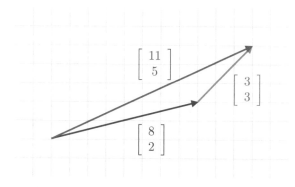

it from other vectors. Any vector $\mathbf{b}$, is transformed into a unit vector $\hat{\mathbf{a}}$, by dividing by its magnitude:

$$\mathbf{b} = [1 \quad 2 \quad 3], \quad ||\mathbf{b}|| = \sqrt{14}, \quad \hat{\mathbf{a}} = \tfrac{1}{\sqrt{14}}[1 \quad 2 \quad 3].$$

3.4.10 Null Vector

A *null vector* has a magnitude of zero and represented by $\mathbf{0}$. Thus, $\mathbf{a}$ and $\mathbf{b}$ are null vectors:

$$\mathbf{a} = [0 \quad 0], \quad \mathbf{b} = [0 \quad 0 \quad 0].$$

3.4.11 Vector Direction

Two- and three-dimensional Euclidean vectors possess magnitude and *direction*, which are encoded within the sign and value of the associated scalars.

3.5 Linearly Dependent and Independent Vectors

Consider the two vectors (3.3):

$$\mathbf{a} = [4 \quad 6], \quad \mathbf{b} = [2 \quad 3] \tag{3.3}$$

one can see that $\mathbf{a}$ is twice $\mathbf{b}$:

$$\mathbf{a} = 2\mathbf{b} \tag{3.4}$$

and are therefore, *linearly dependent*. From (3.4) we observe that $\mathbf{a} - 2\mathbf{b} = \mathbf{0}$, which can be expressed:

$$\lambda_1 \mathbf{a} + \lambda_2 \mathbf{b} = \mathbf{0} \tag{3.5}$$
$$\lambda_1 = 1$$
$$\lambda_2 = -2.$$

Now the two vectors (3.6):

$$\mathbf{a} = [2 \quad 0], \quad \mathbf{b} = [0 \quad 6] \tag{3.6}$$

are *linearly independent*, as the zero terms prevent the vectors from being related to one another. Consequently, using (3.5), both λ_1 and λ_2 must be zero:

$$\lambda_1 \mathbf{a} + \lambda_2 \mathbf{b} = \mathbf{0} \qquad (3.7)$$
$$\lambda_1 = 0$$
$$\lambda_2 = 0.$$

Generalising these observations for any set of vectors, we have:

> A set of vectors $\{\mathbf{v}_1, \mathbf{v}_2, \ldots, \mathbf{v}_n\}$, $n \in \mathbb{N}$, in a vector space V
> is **linearly dependent** if the *only* values of the scalars
> $\lambda_1, \lambda_2, \ldots, \lambda_n \in \mathbb{R}$, are *not* all zero, such that
> $\lambda_1 \mathbf{v}_1 + \lambda_2 \mathbf{v}_2 + \cdots + \lambda_n \mathbf{v}_n = \mathbf{0}$.

> The set of vectors is **linearly independent** if the *only* values of the scalars
> $\lambda_1, \lambda_2, \ldots, \lambda_n \in \mathbb{R}$, are all zero, such that
> $\lambda_1 \mathbf{v}_1 + \lambda_2 \mathbf{v}_2 + \cdots + \lambda_n \mathbf{v}_n = \mathbf{0}$.

Let's test these definitions with another vector.

Consider the two vectors (3.8):

$$\mathbf{a} = [3 \quad 6 \quad 9], \quad \mathbf{b} = [1 \quad 2 \quad 3] \qquad (3.8)$$

one can see that $\mathbf{a}$ is three times $\mathbf{b}$:

$$\mathbf{a} = 3\mathbf{b} \qquad (3.9)$$

therefore,

$$\mathbf{a} - 3\mathbf{b} = \mathbf{0}$$

which implies $\lambda_1 = 1$ and $\lambda_2 = -3$.

Another way to detect dependent and independent vectors is with a determinant. For example, using the two vectors (3.3), we create the following determinant:

$$\begin{vmatrix} 4 & 6 \\ 2 & 3 \end{vmatrix} = 0$$

and notice the zero result. This confirms that the vectors are linearly dependent, because if we use two general dependent vectors:

$$\mathbf{a} = [a_1 \quad a_2], \quad \mathbf{b} = [\lambda a_1 \quad \lambda a_2], \quad \lambda \in \mathbb{R}$$

the corresponding determinant equals zero:

$$\begin{vmatrix} a_1 & a_2 \\ \lambda a_1 & \lambda a_2 \end{vmatrix} = \lambda a_1 a_2 - \lambda a_1 a_2 = 0.$$

If the vectors are independent, such as (3.6), the corresponding determinant is not zero:

$$\begin{vmatrix} 2 & 0 \\ 0 & 6 \end{vmatrix} = 2 \times 6 - 0 \times 0 = 12.$$

Now let's consider three linearly dependent 3-D vectors where $\mathbf{c} = \mathbf{a} + \mathbf{b}$:

$$\mathbf{a} = [1 \quad 2 \quad 1], \quad \mathbf{b} = [3 \quad 4 \quad 3], \quad \mathbf{c} = [4 \quad 6 \quad 4].$$

The corresponding determinant is

$$\begin{vmatrix} 1 & 2 & 1 \\ 3 & 4 & 3 \\ 4 & 6 & 4 \end{vmatrix}.$$

Using Sarrus's rule to expand the determinant, we have:

$$\begin{aligned} \begin{vmatrix} 1 & 2 & 1 \\ 3 & 4 & 3 \\ 4 & 6 & 4 \end{vmatrix} &= (1 \times 4 \times 4) + (2 \times 3 \times 4) + (1 \times 3 \times 6) \\ &\quad - (1 \times 3 \times 6) - (2 \times 3 \times 4) - (1 \times 4 \times 4) \\ &= 16 + 24 + 18 - 18 - 24 - 16 \\ &= 0. \end{aligned}$$

And once again, the determinant records a zero value due to the linear dependence. This can be shown to be a general property by using the following vectors:

$$\begin{aligned} \mathbf{a} &= [a_1 \quad a_2 \quad a_3] \\ \mathbf{b} &= [b_1 \quad b_2 \quad b_3] \\ \mathbf{c} &= [\lambda(a_1 + b_1) \quad \lambda(a_2 + b_2) \quad \lambda(a_3 + b_3)], \quad \lambda \in \mathbb{R}. \end{aligned}$$

Using Sarrus's rule to expand the determinant we have:

$$\begin{vmatrix} a_1 & a_2 & a_3 \\ b_1 & b_2 & b_3 \\ \lambda(a_1 + b_1) & \lambda(a_2 + b_2) & \lambda(a_3 + b_3) \end{vmatrix}$$

$$\begin{aligned} &= a_1 b_2 \lambda(a_3 + b_3) + a_2 b_3 \lambda(a_1 + b_1) + a_3 b_1 \lambda(a_2 + b_2) \\ &\quad - a_1 b_3 \lambda(a_2 + b_2) - a_2 b_1 \lambda(a_3 + b_3) - a_3 b_2 \lambda(a_1 + b_1) \\ &= \lambda(a_1 a_3 b_2 + a_1 b_2 b_3 + a_1 a_2 b_3 + a_2 b_1 b_3 + a_2 a_3 b_1 + a_3 b_1 b_2) \\ &\quad - \lambda(a_1 a_2 b_3 - a_1 b_2 b_3 - a_2 a_3 b_1 - a_2 b_1 b_3 - a_1 a_3 b_2 - a_3 b_1 b_2) \\ &= 0. \end{aligned}$$

The zero value confirms the linear dependence.

Now let's define three linearly independent vectors (3.10):

$$\mathbf{a} = [1 \quad 2 \quad 0], \quad \mathbf{b} = [0 \quad 3 \quad 3], \quad \mathbf{c} = [5 \quad 0 \quad 4]. \tag{3.10}$$

The corresponding determinant is

$$\begin{vmatrix} 1 & 2 & 0 \\ 0 & 3 & 3 \\ 5 & 0 & 4 \end{vmatrix}.$$

Using Sarrus's rule to expand the determinant, we have:

$$\begin{vmatrix} 1 & 2 & 0 \\ 0 & 3 & 3 \\ 5 & 0 & 4 \end{vmatrix} = (1 \times 3 \times 4) + (2 \times 3 \times 5) + (0 \times 0 \times 0)$$

$$- (1 \times 3 \times 0) - (2 \times 0 \times 4) - (0 \times 3 \times 5)$$
$$= 12 + 30 + 0 - 0 - 0 - 0$$
$$= 42.$$

This time, the determinant gives a non-zero value, confirming the independence of the vectors.

3.6 Span of a Set of Vectors

The *span* of a set of vectors tells us something about the space occupied by a linear combination of the vectors. For example, if two vectors are linearly dependent, any linear combination creates a third vector collinear with them. Consequently, their span is confined to an infinite line. To illustrate this, consider the set of vectors S (3.11), where the second vector is 3 times the first vector:

$$S = \left\{ \begin{bmatrix} 3 \\ 1 \end{bmatrix}, \begin{bmatrix} 9 \\ 3 \end{bmatrix} \right\}. \tag{3.11}$$

As the two vectors are linearly related, no matter how they are linearly combined, the span(S) is a line containing the two vectors. See Fig. 3.2.

Fig. 3.2 The two linearly
related vectors

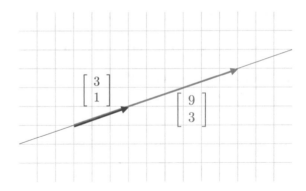

Next, consider the set S of two linearly independent vectors:

$$S = \left\{ \begin{bmatrix} 3 \\ 2 \end{bmatrix}, \begin{bmatrix} 5 \\ 0 \end{bmatrix} \right\}.$$

No matter how one tries, it is impossible to relate the second vector to the first vector using a linear scaling factor. The vectors are linearly independent, and therefore the span(S) is any vector in $\mathbb{R}^2$.

Naturally, only two linearly independent vectors are necessary to span $\mathbb{R}^2$; any extra vector must be a linear combination of the original two. For example, the vector (3.12):

$$\begin{bmatrix} -7 \\ 2 \end{bmatrix} \tag{3.12}$$

appears to be completely unrelated to the two vectors in S, but is related as follows:

$$\begin{bmatrix} -7 \\ 2 \end{bmatrix} = 1 \begin{bmatrix} 3 \\ 2 \end{bmatrix} - 2 \begin{bmatrix} 5 \\ 0 \end{bmatrix}.$$

See Fig. 3.3.

Now consider three vectors in $\mathbb{R}^3$, where there exists some linear dependence. For example, in the following set S, the third vector is the sum of the first two:

$$S = \left\{ \begin{bmatrix} 3 \\ 2 \\ 1 \end{bmatrix}, \begin{bmatrix} 5 \\ 0 \\ 2 \end{bmatrix}, \begin{bmatrix} 8 \\ 2 \\ 3 \end{bmatrix} \right\}.$$

The zero term in the second vector ensures that it is independent of the first vector. Consequently, they coexist on a plane in $\mathbb{R}^3$, and as the third vector is the sum of the first two, it, too, resides on the same plane. This means that the span(S) is a plane in $\mathbb{R}^3$.

Fig. 3.3 How $[-7 \quad 2]^T$ is related to $[3 \quad 2]^T$ and $[5 \quad 0]^T$

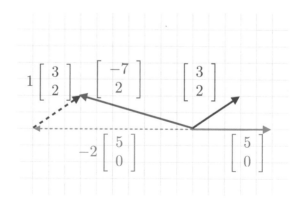

If the set S contains three independent vectors:

$$S = \left\{ \begin{bmatrix} 2 \\ 0 \\ 0 \end{bmatrix}, \begin{bmatrix} 0 \\ 3 \\ 0 \end{bmatrix}, \begin{bmatrix} 0 \\ 0 \\ 4 \end{bmatrix} \right\}$$

then any vector in $\mathbb{R}^3$ can be defined by linearly combining these vectors. Here are two such vectors:

$$\begin{bmatrix} 10 \\ 10 \\ 10 \end{bmatrix} = 5 \begin{bmatrix} 2 \\ 0 \\ 0 \end{bmatrix} + \frac{10}{3} \begin{bmatrix} 0 \\ 3 \\ 0 \end{bmatrix} + \frac{10}{4} \begin{bmatrix} 0 \\ 0 \\ 4 \end{bmatrix}$$

$$\begin{bmatrix} 5 \\ 0 \\ -5 \end{bmatrix} = \frac{5}{2} \begin{bmatrix} 2 \\ 0 \\ 0 \end{bmatrix} + 0 \begin{bmatrix} 0 \\ 3 \\ 0 \end{bmatrix} + \frac{-5}{4} \begin{bmatrix} 0 \\ 0 \\ 4 \end{bmatrix}.$$

Thus we can say that the span$(S) = \mathbb{R}^3$.

3.7 Basis and Dimension

A *basis* is a finite or infinite set of linearly independent vectors that span the associated vector space. A vector is formed by adding together the scaled, unit basis vectors. A simple example of a basis is called the *standard basis*. For two dimensions the basis vectors are $e_1 = [1 \quad 0]$ and $e_2 = [0 \quad 1]$, such that any vector in $\mathbb{R}^2$ is uniquely defined as $a = a_1 e_1 + a_2 e_2$, where $a_1, a_2 \in \mathbb{R}$, and are the vector's Cartesian coordinates. For three dimensions, the basis vectors are $e_1 = [1 \quad 0 \quad 0]$, $e_2 = [0 \quad 1 \quad 0]$ and $e_3 = [0 \quad 0 \quad 1]$, such that any vector in $\mathbb{R}^3$ is uniquely defined as $a = a_1 e_1 + a_2 e_2 + a_3 e_3$. Figure 3.4 shows how a vector in $\mathbb{R}^2$ is formed from the addition of $3e_1$ and $2e_2$.

It is also possible to employ basis vectors that are non-orthogonal.

Fig. 3.4 The role of basis
vectors

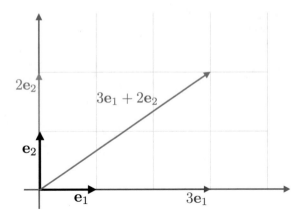

3.7.1   Cartesian Vectors

A *Cartesian vector* is constructed from two or three unit basis vectors: $\mathbf{i}$, $\mathbf{j}$ and $\mathbf{k}$,
aligned with the x-, y- and z-axis, respectively:

$$\mathbf{i} = [1 \quad 0 \quad 0], \qquad \mathbf{j} = [0 \quad 1 \quad 0], \qquad \mathbf{k} = [0 \quad 0 \quad 1].$$

By employing the rules of vector addition and subtraction, we can construct a 2-D
Cartesian vector $\mathbf{a}$ by summing two scaled, unit basis vectors:

$$\mathbf{a} = a_1\mathbf{i} + a_2\mathbf{j}$$

which is equivalent to

$$\mathbf{a} = [a_1 \quad a_2].$$

A 3-D Cartesian vector takes the form:

$$\mathbf{a} = a_1\mathbf{i} + a_2\mathbf{j} + a_3\mathbf{k}$$

and is equivalent to

$$\mathbf{a} = [a_1 \quad a_2 \quad a_3].$$

Any pair of Cartesian vectors, such as $\mathbf{a}$ and $\mathbf{b}$, are combined as follows

$$\mathbf{a} = a_1\mathbf{i} + a_2\mathbf{j} + a_3\mathbf{k}$$
$$\mathbf{b} = b_1\mathbf{i} + b_2\mathbf{j} + b_3\mathbf{k}$$
$$\mathbf{a} \pm \mathbf{b} = (a_1 \pm b_1)\mathbf{i} + (a_2 \pm b_2)\mathbf{j} + (a_3 \pm b_3)\mathbf{k}.$$

3.8 Summary

This chapter begins with a review of the mathematical structures: groups, rings and fields, as these are employed when describing vectors abstractly. In the section on space, the main ideas to carry forward are vector space and a normed vector space. As the reader has probably encountered vectors before, the section on vector definition reviews some important ideas such as vector addition, unit vector and linearly dependent and independent vectors. The determinant is a very useful object in detecting linear dependency and independency. The span of a set of vectors is an important concept, so, too, are basis vectors which lead to the last topic of Cartesian vectors.

References

1. Rovelli C (2014) Reality is not what it seems. Allen Lane. ISBN 978-0-241- 25796-8
2. Petkov V (2009) Relativity and the nature of spacetime. Springer. ISBN 978-3-642-01952-4
3. https://en.wikibooks.org/wiki/Geometry/Five_Postulates_of_Euclidean_Geometry

Chapter 4
Products of Vectors

4.1 Introduction

This chapter describes the scalar and vector products that emerged from Hamilton's quaternions. The algebraic and geometric definitions are explained, together with their associative, distributive and commutative properties. Lastly, the products are combined to form the triple products using geometric analogues.

4.2 Products

We begin with the origin of the two products associated with today's vectorial system: the scalar and vector products. Readers wishing to explore the alternative system of geometric algebra, can refer to the author's books [1] [2].

The scalar and vector products are revealed by examining the product of two Cartesian vectors using a vector function $\circ$, to be defined:

$$\mathbf{a} = a_1\mathbf{i} + a_2\mathbf{j} + a_3\mathbf{k}$$
$$\mathbf{b} = b_1\mathbf{i} + b_2\mathbf{j} + b_3\mathbf{k}$$
$$\mathbf{a} \circ \mathbf{b} = (a_1\mathbf{i} + a_2\mathbf{j} + a_3\mathbf{k}) \circ (b_1\mathbf{i} + b_2\mathbf{j} + b_3\mathbf{k})$$
$$= a_1b_1\mathbf{i} \circ \mathbf{i} + a_1b_2\mathbf{i} \circ \mathbf{j} + a_1b_3\mathbf{i} \circ \mathbf{k}$$
$$+ a_2b_1\mathbf{j} \circ \mathbf{i} + a_2b_2\mathbf{j} \circ \mathbf{j} + a_2b_3\mathbf{j} \circ \mathbf{k}$$
$$+ a_3b_1\mathbf{k} \circ \mathbf{i} + a_3b_2\mathbf{k} \circ \mathbf{j} + a_3b_3\mathbf{k} \circ \mathbf{k}$$
$$= a_1b_1\mathbf{i} \circ \mathbf{i} + a_2b_2\mathbf{j} \circ \mathbf{j} + a_3b_3\mathbf{k} \circ \mathbf{k} \qquad (4.1)$$
$$+ a_1b_2\mathbf{i} \circ \mathbf{j} + a_1b_3\mathbf{i} \circ \mathbf{k}$$
$$+ a_2b_1\mathbf{j} \circ \mathbf{i} + a_2b_3\mathbf{j} \circ \mathbf{k}$$
$$+ a_3b_1\mathbf{k} \circ \mathbf{i} + a_3b_2\mathbf{k} \circ \mathbf{j}.$$

© Springer-Verlag London Ltd., part of Springer Nature 2021
J. Vince, *Vector Analysis for Computer Graphics*,
https://doi.org/10.1007/978-1-4471-7505-6_4

Equation (4.1) contains two parts that can be separated into scalar and vector portions, which is effected by two different functions acting upon the Cartesian vectors. If we make $\circ = \cdot$, where $\mathbf{a} \cdot \mathbf{b} = ||\mathbf{a}||\,||\mathbf{b}||\cos\theta$, and θ is the angle between $\mathbf{a}$ and $\mathbf{b}$, then

$$\mathbf{a} \cdot \mathbf{b} = a_1 b_1 \mathbf{i} \cdot \mathbf{i} + a_2 b_2 \mathbf{j} \cdot \mathbf{j} + a_3 b_3 \mathbf{k} \cdot \mathbf{k}$$
$$+ a_1 b_2 \mathbf{i} \cdot \mathbf{j} + a_1 b_3 \mathbf{i} \cdot \mathbf{k}$$
$$+ a_2 b_1 \mathbf{j} \cdot \mathbf{i} + a_2 b_3 \mathbf{j} \cdot \mathbf{k}$$
$$+ a_3 b_1 \mathbf{k} \cdot \mathbf{i} + a_3 b_2 \mathbf{k} \cdot \mathbf{j}.$$

The products $\mathbf{i} \cdot \mathbf{i}$, $\mathbf{j} \cdot \mathbf{j}$ and $\mathbf{k} \cdot \mathbf{k}$ equal 1, due to their mutual alignment, and the remaining vector products equal zero, due to their orthogonality. Thus:

$$\mathbf{a} \cdot \mathbf{b} = a_1 b_1 + a_2 b_2 + a_3 b_3$$

creating a scalar.

Conversely, if we make $\circ = \times$, with $\mathbf{a} \times \mathbf{b} = ||\mathbf{a}||\,||\mathbf{b}||\sin\theta\,\hat{\mathbf{n}}$, where $\hat{\mathbf{n}}$ is a unit vector perpendicular to the plane containing $\mathbf{a}$ and $\mathbf{b}$, then

$$\mathbf{a} \times \mathbf{b} = a_1 b_1 \mathbf{i} \times \mathbf{i} + a_2 b_2 \mathbf{j} \times \mathbf{j} + a_3 b_3 \mathbf{k} \times \mathbf{k}$$
$$+ a_1 b_2 \mathbf{i} \times \mathbf{j} + a_1 b_3 \mathbf{i} \times \mathbf{k}$$
$$+ a_2 b_1 \mathbf{j} \times \mathbf{i} + a_2 b_3 \mathbf{j} \times \mathbf{k}$$
$$+ a_3 b_1 \mathbf{k} \times \mathbf{i} + a_3 b_2 \mathbf{k} \times \mathbf{j}.$$

The products $\mathbf{i} \times \mathbf{i}$, $\mathbf{j} \times \mathbf{j}$ and $\mathbf{k} \times \mathbf{k}$ equal 0, due to their mutual alignment, and the remaining terms equal 1, due to their orthogonality. Thus:

$$\mathbf{a} \times \mathbf{b} = a_1 b_2 \mathbf{i} \times \mathbf{j} + a_1 b_3 \mathbf{i} \times \mathbf{k} + a_2 b_1 \mathbf{j} \times \mathbf{i} + a_2 b_3 \mathbf{j} \times \mathbf{k} + a_3 b_1 \mathbf{k} \times \mathbf{i} + a_3 b_2 \mathbf{k} \times \mathbf{j}.$$

Hamilton's work on quaternions reveal the loss of commutativity: $\mathbf{j} \times \mathbf{i} = -\mathbf{i} \times \mathbf{j}$, etc., and the cyclic relationship between $\mathbf{i}$, $\mathbf{j}$ and $\mathbf{k}$:

$$\mathbf{i} \times \mathbf{j} = \mathbf{k}, \quad \mathbf{j} \times \mathbf{k} = \mathbf{i}, \quad \mathbf{k} \times \mathbf{i} = \mathbf{j}.$$

Therefore,

$$\mathbf{a} \times \mathbf{b} = (a_1 b_2 - a_2 b_1)\mathbf{i} \times \mathbf{j} + (a_3 b_1 - a_1 b_3)\mathbf{k} \times \mathbf{i} + (a_2 b_3 - a_3 b_2)\mathbf{j} \times \mathbf{k}$$
$$= (a_1 b_2 - a_2 b_1)\mathbf{k} + (a_3 b_1 - a_1 b_3)\mathbf{j} + (a_2 b_3 - a_3 b_2)\mathbf{i}$$
$$= (a_2 b_3 - a_3 b_2)\mathbf{i} + (a_3 b_1 - a_1 b_3)\mathbf{j} + (a_1 b_2 - a_2 b_1)\mathbf{k}$$

creating a vector. The use of the '×' symbol for the product has created the alternative name: *cross product*.

It's worth pointing out that this vector product does not exist in geometric algebra – Hamilton's quaternions play no part. Instead, the vector product is replaced by the *wedge product*, which interprets the components of the vector as signed areas. The wedge product makes $\circ = \wedge$, where $\mathbf{a} \wedge \mathbf{b} = ||\mathbf{a}|| \, ||\mathbf{b}|| \, \sin\theta$, then,

$$\mathbf{a} \wedge \mathbf{b} = (a_2b_3 - a_3b_2)\mathbf{j} \wedge \mathbf{k} + (a_3b_1 - a_1b_3)\mathbf{k} \wedge \mathbf{i} + (a_1b_2 - a_2b_1)\mathbf{i} \wedge \mathbf{j}$$

and $\mathbf{j} \wedge \mathbf{k}, \quad \mathbf{k} \wedge \mathbf{i}$ and $\mathbf{i} \wedge \mathbf{j}$ are called *bivectors* that are signed areas.

Let us now consider these two products individually.

4.3 Scalar Product

4.3.1 Algebraic Definition

The *scalar product* of two vectors of the same dimension is defined as the sum of the products of the corresponding elements. Given two vectors $\mathbf{a}$ and $\mathbf{b}$:

$$\mathbf{a} = [a_1 \quad a_2 \quad \cdots \quad a_n], \quad \mathbf{b} = [b_1 \quad b_2 \quad \cdots \quad b_n], \quad \mathbf{a} \cdot \mathbf{b} = a_1b_1 + a_2b_2 + \cdots + a_nb_n$$

or more succinctly:

$$\sum_{i=1}^{n} a_ib_i.$$

The scalar product is also known as the *dot product* due to the 'dot' representing the multiplication symbol. For example, given two vectors $\mathbf{a}$ and $\mathbf{b}$:

$$\mathbf{a} = [2 \quad 3 \quad 4], \quad \mathbf{b} = [6 \quad 7 \quad 8], \quad \mathbf{a} \cdot \mathbf{b} = 2 \cdot 6 + 3 \cdot 7 + 4 \cdot 8 = 65.$$

4.3.2 Axioms

As the dot product is just the sum of scalar products, it satisfies the commutative multiplication, and distributivity axioms:

Commutative multiplication: $\mathbf{a} \cdot \mathbf{b} = \mathbf{b} \cdot \mathbf{a}$.

Distributivity: $\mathbf{a} \cdot (\mathbf{b} + \mathbf{c}) = \mathbf{a} \cdot \mathbf{b} + \mathbf{a} \cdot \mathbf{c}$

$$\mathbf{a} \cdot \mathbf{a} = ||\mathbf{a}||^2$$

$$\mathbf{a} \cdot \mathbf{0} = \mathbf{0}$$

$$(\lambda\mathbf{a}) \cdot \mathbf{b}) = \mathbf{a} \cdot (\lambda\mathbf{b}) = \lambda(\mathbf{a} \cdot \mathbf{b}), \quad \lambda \in \mathbb{R}.$$

4.3.3 Geometric Definition

The dot product also has a geometric definition:

$$\mathbf{a} \cdot \mathbf{b} = ||\mathbf{a}|| \, ||\mathbf{b}|| \cos \theta$$

which is proved as follows.

Figure 4.1 shows two vectors $\mathbf{a}$ and $\mathbf{b}$ separated by an angle θ, and their difference vector $\mathbf{a} - \mathbf{b}$. The axioms of the scalar product inform us that:

$$\mathbf{a} = [a_1 \quad a_2 \quad \cdots \quad a_n]$$
$$||\mathbf{a}||^2 = a_1^2 + a_2^2 + \cdots + a_n^2 = \mathbf{a} \cdot \mathbf{a}.$$

Therefore,

$$\begin{aligned}
||\mathbf{a} - \mathbf{b}||^2 &= (\mathbf{a} - \mathbf{b}) \cdot (\mathbf{a} - \mathbf{b}) \\
&= \mathbf{a} \cdot \mathbf{a} - \mathbf{a} \cdot \mathbf{b} - \mathbf{b} \cdot \mathbf{a} + \mathbf{b} \cdot \mathbf{b} \\
&= \mathbf{a} \cdot \mathbf{a} - 2\mathbf{a} \cdot \mathbf{b} + \mathbf{b} \cdot \mathbf{b} \\
&= ||\mathbf{a}||^2 + ||\mathbf{b}||^2 - 2\mathbf{a} \cdot \mathbf{b}.
\end{aligned} \tag{4.2}$$

Applying the cosine rule to the triangle in Fig. 4.1 we have:

$$||\mathbf{a} - \mathbf{b}||^2 = ||\mathbf{a}||^2 + ||\mathbf{b}||^2 - 2||\mathbf{a}|| \, ||\mathbf{b}|| \cos \theta. \tag{4.3}$$

Substituting (4.2) in (4.3) we have:

$$\begin{aligned}
||\mathbf{a}||^2 + ||\mathbf{b}||^2 - 2\mathbf{a} \cdot \mathbf{b} &= ||\mathbf{a}||^2 + ||\mathbf{b}||^2 - 2||\mathbf{a}|| \, ||\mathbf{b}|| \cos \theta \\
-2\mathbf{a} \cdot \mathbf{b} &= -2||\mathbf{a}|| \, ||\mathbf{b}|| \cos \theta \\
\mathbf{a} \cdot \mathbf{b} &= ||\mathbf{a}|| \, ||\mathbf{b}|| \cos \theta.
\end{aligned}$$

Fig. 4.1 Two vectors $\mathbf{a}$ and $\mathbf{b}$ and their difference vector $\mathbf{a} - \mathbf{b}$

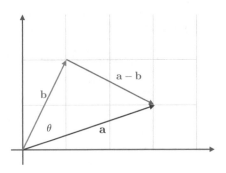

We can quickly test this relationship with a simple example. Given two unit vectors $\hat{\mathbf{a}} = 1\mathbf{i} + 0\mathbf{j}$ and $\hat{\mathbf{b}} = \frac{\sqrt{2}}{2}\mathbf{i} + \frac{\sqrt{2}}{2}\mathbf{j}$, separated by an angle of $\theta = \pi/4$ radians (45°), then:

$$\hat{\mathbf{a}} \cdot \hat{\mathbf{b}} = \begin{bmatrix} 1 & 0 \end{bmatrix} \cdot \begin{bmatrix} \frac{\sqrt{2}}{2} \\ \frac{\sqrt{2}}{2} \end{bmatrix} = \frac{\sqrt{2}}{2}$$

$$\hat{\mathbf{a}} \cdot \hat{\mathbf{b}} = ||\hat{\mathbf{a}}|| \, ||\hat{\mathbf{b}}|| \cos\theta = \cos\left(\frac{\pi}{4}\right) = \frac{\sqrt{2}}{2}.$$

4.4 Vector Product

The *vector product* of two vectors, creates a third vector orthogonal to the plane containing the vectors, and is very useful in calculating surface normals. It is very similar to Hamilton's quaternions where the imaginary units i, j, k are related by the non-commutative properties: $ij = k, \ jk = i, \ ki = j$.

Both the scalar and the vector products were independently introduced by Josiah Willard Gibbs and the English electrical engineer and mathematician Oliver Heaviside [1850–1925], who published in 1893 his description of vector algebra and analysis [3].

Oliver Heaviside was no supporter of quaternions, as this quotation from his book shows:

> But that was before I had thrown off the quaternionic old-man-of-the sea who fastened himself on my shoulders when reading the only accessible treatise on the subject–Prof. Tait's Quaternions. But I came later to see that, so far as the vector analysis I required was concerned, the quaternion was not only not required, but was a positive evil of no inconsiderable magnitude; and that by its avoidance the establishment of vector analysis was made quite simple and its working also simplified, and that it could be conveniently harmonised with ordinary Cartesian work.

Heaviside defined the vector product, using the word tensor to mean modulus, as:

> But *the* vector product of two vectors has a strictly-fixed tensor, depending upon those of the component vectors and their inclination. There is particular advantage in taking the tensor of the vector product of **a** and **b** to be $ab \sin\theta$. Thus we *define* the vector product of two vectors **a** and **b** whose tensors are a and b, and whose included angle is θ, to be a third vector **c** whose tensor c equals $ab \sin\theta$, and whose direction is perpendicular to the plane of **a** and **b**; the positive direction of **c** being such that positive or right-handed rotation about **c** carries the vector **a** to **b**. This vector product is denoted by

$$\mathbf{c} = \mathbf{Vab},$$

> and its tensor may be denoted by $\mathbf{V_0 ab} = ab \sin\theta$.

Heaviside employed **V** to represent the vector product, which today has been replaced by ×. He continues to show that

$$\mathbf{i} \times \mathbf{j} = \mathbf{k}, \quad \mathbf{j} \times \mathbf{k} = \mathbf{i}, \quad \mathbf{k} \times \mathbf{i} = \mathbf{j}$$

$$\mathbf{i} \times \mathbf{i} = 0, \quad \mathbf{j} \times \mathbf{j} = 0, \quad \mathbf{k} \times \mathbf{k} = 0.$$

Lastly, he writes:

Since the tensor of $\mathbf{Vab}$ is $ab \sin \theta$ and the scalar product $\mathbf{ab}$ is $ab \cos \theta$, we have

$$(\mathbf{ab})^2 + (\mathbf{Vab})^2 = (ab)^2 .$$

This last statement is a version of the *Lagrange identity* invented by the Franco-Italian mathematician and astronomer Joseph-Louis Lagrange [1736–1813], which in vector calculus links the dot product, vector product with the lengths of the associated vectors.

Given two vectors $\mathbf{a}$ and $\mathbf{b}$, their vector product is:

$$\mathbf{a} = a_1\mathbf{i} + a_2\mathbf{j} + a_3\mathbf{k}$$
$$\mathbf{b} = b_1\mathbf{i} + b_2\mathbf{j} + b_3\mathbf{k}$$
$$\mathbf{a} \times \mathbf{b} = (a_2b_3 - a_3b_2)\mathbf{i} + (a_3b_1 - a_1b_3)\mathbf{j} + (a_1b_2 - a_2b_1)\mathbf{k}.$$

An *aide-mémoire* for computing this relationship uses the following determinant, with the $\mathbf{j}$ component reversed:

$$\mathbf{a} \times \mathbf{b} = \begin{vmatrix} \mathbf{i} & \mathbf{j} & \mathbf{k} \\ a_1 & a_2 & a_3 \\ b_1 & b_2 & b_3 \end{vmatrix} = \begin{vmatrix} a_2 & a_3 \\ b_2 & b_3 \end{vmatrix} \mathbf{i} - \begin{vmatrix} a_1 & a_3 \\ b_1 & b_3 \end{vmatrix} \mathbf{j} + \begin{vmatrix} a_1 & a_2 \\ b_1 & b_2 \end{vmatrix} \mathbf{k}$$

$$\mathbf{a} \times \mathbf{b} = (a_2b_3 - a_3b_2)\mathbf{i} - (a_1b_3 - a_3b_1)\mathbf{j} + (a_1b_2 - a_2b_1)\mathbf{k}.$$

Now let's prove $||\mathbf{a} \times \mathbf{b}|| = ||\mathbf{a}|| \, ||\mathbf{b}|| \, |\sin \theta|$:

$$||\mathbf{a}||^2 = \mathbf{a} \cdot \mathbf{a}$$
$$||\mathbf{a} \times \mathbf{b}||^2 = (\mathbf{a} \times \mathbf{b}) \cdot (\mathbf{a} \times \mathbf{b})$$
$$= (a_2b_3 - a_3b_2)^2 + (a_1b_3 - a_3b_1)^2 + (a_1b_2 - a_2b_1)^2$$
$$= a_1^2 \left(b_2^2 + b_3^2\right) + a_2^2 \left(b_3^2 + b_1^2\right) + a_3^2 \left(b_2^2 + b_1^2\right)$$
$$- 2(a_2a_3b_2b_3 + a_1a_3b_1b_3 + a_1a_2b_1b_2)$$
$$(\mathbf{a} \cdot \mathbf{b})^2 = (a_1b_1 + a_2b_2 + a_3b_3)^2$$
$$= a_1^2b_1^2 + a_2^2b_2^2 + a_3^2b_3^2 + 2(a_1a_2b_1b_2 + a_1a_3b_1b_3 + a_2a_3b_2b_3)$$
$$||\mathbf{a} \times \mathbf{b}||^2 + (\mathbf{a} \cdot \mathbf{b})^2 = a_1^2 \left(b_1^2 + b_2^2 + b_3^2\right) + a_2^2 \left(b_1^2 + b_2^2 + b_3^2\right) + a_3^2 \left(b_1^2 + b_2^2 + b_3^2\right)$$
$$= \left(a_1^2 + a_2^2 + a_3^2\right)\left(b_1^2 + b_2^2 + b_3^2\right)$$
$$= ||\mathbf{a}||^2 ||\mathbf{b}||^2$$
$$||\mathbf{a} \times \mathbf{b}||^2 = ||\mathbf{a}||^2 ||\mathbf{b}||^2 - (\mathbf{a} \cdot \mathbf{b})^2$$

$$= ||\mathbf{a}||^2||\mathbf{b}||^2 - ||\mathbf{a}||^2||\mathbf{b}||^2 \cos^2 \theta$$
$$= ||\mathbf{a}||^2||\mathbf{b}||^2 \left(1 - \cos^2 \theta\right)$$
$$= ||\mathbf{a}||^2||\mathbf{b}||^2 \sin^2 \theta$$
$$||\mathbf{a} \times \mathbf{b}|| = ||\mathbf{a}||\,||\mathbf{b}||\,|\sin \theta|.$$

Let's test the vector product with the following scenario:

$$\mathbf{a} = 3\mathbf{i} + 1\mathbf{j} + 0\mathbf{k}$$
$$\mathbf{b} = 1\mathbf{i} + 2\mathbf{j} + 0\mathbf{k}$$

$$\mathbf{a} \times \mathbf{b} = \begin{vmatrix} \mathbf{i} & \mathbf{j} & \mathbf{k} \\ 3 & 1 & 0 \\ 1 & 2 & 0 \end{vmatrix} = \begin{vmatrix} 1 & 0 \\ 2 & 0 \end{vmatrix} \mathbf{i} - \begin{vmatrix} 3 & 0 \\ 1 & 0 \end{vmatrix} \mathbf{j} + \begin{vmatrix} 3 & 1 \\ 1 & 2 \end{vmatrix} \mathbf{k}$$

$$= 0\mathbf{i} + 0\mathbf{j} + 5\mathbf{k}.$$

Which is correct, as $\mathbf{a}$ and $\mathbf{b}$ are confined to the $\mathbf{ij}$ plane. Let's use the scalar product to determine the angle between $\mathbf{a}$ and $\mathbf{b}$, and substitute this back into the vector product:

$$||\mathbf{a}|| = \sqrt{3^2 + 1^2} = \sqrt{10}$$
$$||\mathbf{b}|| = \sqrt{1^2 + 2^2} = \sqrt{5}$$
$$\mathbf{a} \cdot \mathbf{b} = 3 \cdot 1 + 1 \cdot 2 = 5$$
$$\mathbf{a} \cdot \mathbf{b} = ||\mathbf{a}||\,||\mathbf{b}||\cos \theta$$
$$\theta = \cos^{-1} \left(\frac{\mathbf{a} \cdot \mathbf{b}}{||\mathbf{a}||\,||\mathbf{b}||} \right)$$
$$= \cos^{-1} \left(\frac{5}{\sqrt{10}\sqrt{5}} \right) = 45°$$
$$||\mathbf{a} \times \mathbf{b}|| = ||\mathbf{a}||\,||\mathbf{b}||\,|\sin \theta|$$
$$= \sqrt{10}\sqrt{5} \sin 45°$$
$$= 5.$$

Now let's demonstrate the non-commutativity of the vector product by reversing the vectors $\mathbf{a}$ and $\mathbf{b}$:

$$\mathbf{a} = 1\mathbf{i} + 2\mathbf{j} + 0\mathbf{k}$$
$$\mathbf{b} = 3\mathbf{i} + 1\mathbf{j} + 0\mathbf{k}$$

$$\mathbf{a} \times \mathbf{b} = \begin{vmatrix} \mathbf{i} & \mathbf{j} & \mathbf{k} \\ 1 & 2 & 0 \\ 3 & 1 & 0 \end{vmatrix} = \begin{vmatrix} 2 & 0 \\ 1 & 0 \end{vmatrix} \mathbf{i} - \begin{vmatrix} 1 & 0 \\ 3 & 0 \end{vmatrix} \mathbf{j} + \begin{vmatrix} 1 & 2 \\ 3 & 1 \end{vmatrix} \mathbf{k}$$

$$= 0\mathbf{i} + 0\mathbf{j} - 5\mathbf{k}.$$

Which confirms that the vector has been reversed.

4.4.1   Geometric Definition

The modulus of the vector product: $||\mathbf{a} \times \mathbf{b}||$ equals $||\mathbf{a}|| \, ||\mathbf{b}|| \, |\sin\theta|$, which is the area of the parallelogram that $\mathbf{a}$ and $\mathbf{b}$ span, as shown in Fig. 4.2. It is this relationship that is behind the bivector in geometric algebra.

4.4.2   The Right-Hand Rule

The *right-hand rule* identifies the directions of the three vectors involved in the vector product. The vector product is defined as

$$\mathbf{a} \times \mathbf{b} = ||\mathbf{a}|| \, ||\mathbf{b}|| \, \sin\theta \, \hat{\mathbf{n}}$$

with vectors $\mathbf{a}$ and $\mathbf{b}$ separated by an angle θ, and $\hat{\mathbf{n}}$ is perpendicular to the plane containing $\mathbf{a}$ and $\mathbf{b}$. And because sin is an odd function, i.e. $\sin-\theta = -\sin\theta$, the direction of $\mathbf{a} \times \mathbf{b}$ is reversed by changing the sign of θ.

Therefore, with a positive θ, and using one's right hand, the index (first) finger points along $\mathbf{a}$, the middle finger points along $\mathbf{b}$, which makes the thumb point in the direction of $\hat{\mathbf{n}}$, i.e. $\mathbf{a} \times \mathbf{b}$. As a test let $\mathbf{a} = \mathbf{i}$, $\mathbf{b} = \mathbf{j}$, and the thumb points along $\mathbf{k}$. If vectors $\mathbf{a}$ and $\mathbf{b}$ are reversed, $\hat{\mathbf{n}}$ is reversed. However, all of this is taken care of by the determinant:

$$\mathbf{a} = 1\mathbf{i} + 0\mathbf{j} + 0\mathbf{k}$$
$$\mathbf{b} = 0\mathbf{i} + 1\mathbf{j} + 0\mathbf{k}$$

$$\mathbf{a} \times \mathbf{b} = \begin{vmatrix} \mathbf{i} & \mathbf{j} & \mathbf{k} \\ 1 & 0 & 0 \\ 0 & 1 & 0 \end{vmatrix} = \begin{vmatrix} 0 & 0 \\ 1 & 0 \end{vmatrix} \mathbf{i} - \begin{vmatrix} 1 & 0 \\ 0 & 0 \end{vmatrix} \mathbf{j} + \begin{vmatrix} 1 & 0 \\ 0 & 1 \end{vmatrix} \mathbf{k}$$

$$= 0\mathbf{i} + 0\mathbf{j} + 1\mathbf{k}.$$

Fig. 4.2 The area created by the vectors **a** and **b**

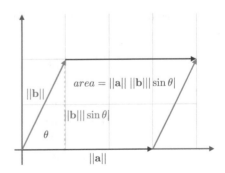

Reversing the vectors:

$$\mathbf{a} = 0\mathbf{i} + 1\mathbf{j} + 0\mathbf{k}$$
$$\mathbf{b} = 1\mathbf{i} + 0\mathbf{j} + 0\mathbf{k}$$

$$\mathbf{a} \times \mathbf{b} = \begin{vmatrix} \mathbf{i} & \mathbf{j} & \mathbf{k} \\ 0 & 1 & 0 \\ 1 & 0 & 0 \end{vmatrix} = \begin{vmatrix} 1 & 0 \\ 0 & 0 \end{vmatrix} \mathbf{i} - \begin{vmatrix} 0 & 0 \\ 1 & 0 \end{vmatrix} \mathbf{j} + \begin{vmatrix} 0 & 1 \\ 1 & 0 \end{vmatrix} \mathbf{k}$$

$$= 0\mathbf{i} + 0\mathbf{j} - 1\mathbf{k}.$$

And the direction is reversed.

4.5 Triple Products

So far, we have examined the scalar and vector products of two vectors; now let's examine two *triple products* of three vectors: the *scalar triple product* and the *vector triple product*.

4.5.1 Scalar Triple Product

The *scalar triple product* computes the volume of a space defined by three vectors $\mathbf{a}$, $\mathbf{b}$ and $\mathbf{c}$, and is written:

$$V = \mathbf{c} \cdot (\mathbf{a} \times \mathbf{b}). \tag{4.4}$$

Figure 4.3 shows three vectors $\mathbf{a}$, $\mathbf{b}$ and $\mathbf{c}$, that provide the basis for a prism, where the area of the base is given by $A = ||\mathbf{a}||\ ||\mathbf{b}|| \sin \theta$. But $\mathbf{n} = \mathbf{a} \times \mathbf{b}$, where $||\mathbf{n}|| = A$. Now the volume of a prism is the product of its perpendicular height and its base area, which is given by

$$V = (||\mathbf{c}||\ \cos \alpha)\,(||\mathbf{n}||) = ||\mathbf{c}||\ ||\mathbf{n}||\ \cos \alpha$$

which is the dot product $\mathbf{c} \cdot \mathbf{n}$. But as $\mathbf{n} = \mathbf{a} \times \mathbf{b}$, we can state:

$$V = \mathbf{c} \cdot (\mathbf{a} \times \mathbf{b}).$$

The prism in Fig. 4.3 could be drawn with vectors $\mathbf{b}$ and $\mathbf{c}$ as the base, which using the above reasoning would make

$$V = \mathbf{a} \cdot (\mathbf{b} \times \mathbf{c}).$$

Fig. 4.3 The prism volume
calculated by the scalar triple
product

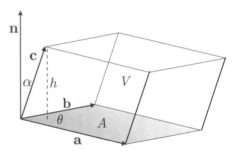

Similarly, if vectors **c** and **a** form the base, the volume is

$$V = \mathbf{b} \cdot (\mathbf{c} \times \mathbf{a}).$$

Thus, we have:

$$V = \mathbf{a} \cdot (\mathbf{b} \times \mathbf{c}) = \mathbf{b} \cdot (\mathbf{c} \times \mathbf{a}) = \mathbf{c} \cdot (\mathbf{a} \times \mathbf{b}).$$

Given:

$$\mathbf{a} = a_1\mathbf{i} + a_2\mathbf{j} + a_3\mathbf{k}$$
$$\mathbf{b} = b_1\mathbf{i} + b_2\mathbf{j} + b_3\mathbf{k}$$
$$\mathbf{c} = c_1\mathbf{i} + c_2\mathbf{j} + c_3\mathbf{k}$$
$$\mathbf{a} \times \mathbf{b} = (a_2b_3 - a_3b_2)\mathbf{i} - (a_1b_3 - a_3b_1)\mathbf{j} + (a_1b_2 - a_2b_1)\mathbf{k}$$
$$\mathbf{c} \cdot (\mathbf{a} \times \mathbf{b}) = (c_1\mathbf{i} + c_2\mathbf{j} + c_3\mathbf{k}) \cdot [(a_2b_3 - a_3b_2)\mathbf{i} - (a_1b_3 - a_3b_1)\mathbf{j} + (a_1b_2 - a_2b_1)\mathbf{k}]$$
$$= (a_2b_3 - a_3b_2)c_1 - (a_1b_3 - a_3b_1)c_2 + (a_1b_2 - a_2b_1)c_3$$
$$= a_1b_2c_3 + a_2b_3c_1 + a_3b_1c_2 - a_1b_3c_2 - a_2b_1c_3 - a_3b_2c_1$$

which is the Sarrus expansion of a third-order determinant:

$$\mathbf{c} \cdot (\mathbf{a} \times \mathbf{b}) = \begin{vmatrix} a_1 & a_2 & a_3 \\ b_1 & b_2 & b_3 \\ c_1 & c_2 & c_3 \end{vmatrix}. \qquad (4.5)$$

Let's test (4.5) with a 2-unit sided cube:

$$\mathbf{a} = 2\mathbf{i}, \quad \mathbf{b} = 2\mathbf{j}, \quad \mathbf{c} = 2\mathbf{k}$$
$$\mathbf{c} \cdot (\mathbf{a} \times \mathbf{b}) = \begin{vmatrix} 2 & 0 & 0 \\ 0 & 2 & 0 \\ 0 & 0 & 2 \end{vmatrix} = 8.$$

4.5.2 Vector Triple Product

The *vector triple product* is written:

$$\mathbf{v} = \mathbf{a} \times (\mathbf{b} \times \mathbf{c}). \tag{4.6}$$

As (4.6) contains two cross products, one of them is enclosed within parentheses to distinguish it from the alternative form:

$$\mathbf{v} = (\mathbf{a} \times \mathbf{b}) \times \mathbf{c}$$

which, due to the non-associative nature of the cross product, would generally give a different result.

If $\mathbf{b}$ and $\mathbf{c}$ in (4.6) are linearly related, then $\mathbf{b} \times \mathbf{c} = \mathbf{0}$, and $\mathbf{v} = \mathbf{0}$. But, if $\mathbf{b}$ and $\mathbf{c}$ are not linearly related, they define a plane containing both vectors, such that $\mathbf{b} \times \mathbf{c}$ is perpendicular to the plane using the right-hand rule. This, in turn, implies that $\mathbf{a} \times (\mathbf{b} \times \mathbf{c})$ is perpendicular to $\mathbf{b} \times \mathbf{c}$, and $\mathbf{a} \times (\mathbf{b} \times \mathbf{c})$ resides in the plane containing $\mathbf{b}$ and $\mathbf{c}$.

For example, Fig. 4.4 shows the following scenario:

$$\mathbf{a} = [0 \ \ 0 \ \ 1], \quad \mathbf{b} = [3 \ \ 0 \ \ 0], \quad \mathbf{c} = [0 \ \ 2 \ \ 2]$$

with the plane containing $\mathbf{b}$ and $\mathbf{c}$ shaded green.

The value of $\mathbf{b} \times \mathbf{c}$ is calculated as follows:

$$\mathbf{b} \times \mathbf{c} = \begin{vmatrix} \mathbf{i} & \mathbf{j} & \mathbf{k} \\ 3 & 0 & 0 \\ 0 & 2 & 2 \end{vmatrix} = \begin{vmatrix} 0 & 0 \\ 2 & 2 \end{vmatrix} \mathbf{i} - \begin{vmatrix} 3 & 0 \\ 0 & 2 \end{vmatrix} \mathbf{j} + \begin{vmatrix} 3 & 0 \\ 0 & 2 \end{vmatrix} \mathbf{k}$$
$$= 0\mathbf{i} - 6\mathbf{j} + 6\mathbf{k}.$$

The value of $\mathbf{a} \times (\mathbf{b} \times \mathbf{c})$ is calculated as follows:

$$\mathbf{a} \times (\mathbf{b} \times \mathbf{c}) = \begin{vmatrix} \mathbf{i} & \mathbf{j} & \mathbf{k} \\ 0 & 0 & 1 \\ 0 & -6 & 6 \end{vmatrix} = \begin{vmatrix} 0 & 1 \\ -6 & 6 \end{vmatrix} \mathbf{i} - \begin{vmatrix} 0 & 1 \\ 0 & 6 \end{vmatrix} \mathbf{j} + \begin{vmatrix} 0 & 0 \\ 0 & -6 \end{vmatrix} \mathbf{k}$$
$$= 6\mathbf{i} + 0\mathbf{j} + 0\mathbf{k}.$$

Thus $\mathbf{a} \times (\mathbf{b} \times \mathbf{c}) = 6\mathbf{i}$, which is in the plane containing $\mathbf{b}$ and $\mathbf{c}$.

Knowing that $\mathbf{b}$ and $\mathbf{c}$ provide a basis that spans the plane containing $\mathbf{b}$ and $\mathbf{c}$, and that $\mathbf{a} \times (\mathbf{b} \times \mathbf{c})$ belongs to this plane, we can state:

$$\mathbf{a} \times (\mathbf{b} \times \mathbf{c}) = \lambda_1 \mathbf{b} + \lambda_2 \mathbf{c}, \quad \lambda_1, \lambda_2 \in \mathbb{R}.$$

In the above example, $\mathbf{a} \times (\mathbf{b} \times \mathbf{c}) = 2\mathbf{b}$.

Fig. 4.4 The vector product
b × **c** and the resulting
vector

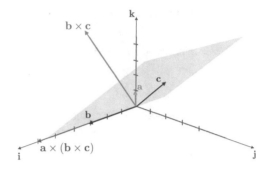

It can be shown that $\lambda_1 = \mathbf{a} \cdot \mathbf{c}$, and $\lambda_2 = -\mathbf{a} \cdot \mathbf{b}$:

$$\mathbf{a} \times (\mathbf{b} \times \mathbf{c}) = (\mathbf{a} \cdot \mathbf{c})\mathbf{b} - (\mathbf{a} \cdot \mathbf{b})\mathbf{c}$$

which in the above example produces:

$$\mathbf{a} \times (\mathbf{b} \times \mathbf{c}) = \left(\begin{bmatrix} 0 \\ 0 \\ 1 \end{bmatrix} \cdot \begin{bmatrix} 0 \\ 2 \\ 2 \end{bmatrix} \right) \begin{bmatrix} 3 \\ 0 \\ 0 \end{bmatrix} - \left(\begin{bmatrix} 0 \\ 0 \\ 1 \end{bmatrix} \cdot \begin{bmatrix} 3 \\ 0 \\ 0 \end{bmatrix} \right) \begin{bmatrix} 0 \\ 2 \\ 2 \end{bmatrix}$$

$$= 2 \begin{bmatrix} 3 \\ 0 \\ 0 \end{bmatrix} - \mathbf{0}$$

$$= \begin{bmatrix} 6 \\ 0 \\ 0 \end{bmatrix}$$

$$= 6\mathbf{i} + 0\mathbf{j} + 0\mathbf{k}.$$

4.6   Summary

This chapter shows the historical development from quaternion to vector products.
The two products associated with vectors are the scalar and vector product, where
the first is used to determine the angle between two vectors, and the second creates a
vector normal to the plane containing the two vectors. The scalar product is associative, whereas the vector product is non-associative, and the right-hand rule is used
to determine the direction of the normal vector.

The chapter finishes with the two triple products associated with vectors, where
the scalar triple product generates the volume of a parallelepiped, and the vector
triple product generates a vector aligned with the plane containing the dominant pair
of vectors.

References

1. Vince J (2009) Geometric algebra for computer graphics. Springer. ISBN 978-1-84628-996-5
2. Vince J (2009) Geometric algebra: an algebraic system for computer games and animation. Springer. ISBN 978-1-84882-378-5
3. Heaviside O (1893) Electromagnetic theory chap. 3, vol 1, pp 132–305

Chapter 5
Differentiating Vector-Valued Functions

5.1 Introduction

This chapter introduces vector-valued functions, and shows how they are differentiated and integrated. In particular, we show how such functions are used to represent velocity and acceleration vectors. The chapter assumes that readers are familiar with calculus. If you need to brush up your calculus, then the following books may be useful [1–3].

5.2 Vector-Valued Functions

A *vector-valued function* takes the form:

$$\mathbf{p}(t) = x(t)\mathbf{i}$$
$$\mathbf{p}(t) = x(t)\mathbf{i} + y(t)\mathbf{j}$$
$$\mathbf{p}(t) = x(t)\mathbf{i} + y(t)\mathbf{j} + z(t)\mathbf{k}$$

etc.

where $\mathbf{p}(t)$ is a vector, and each basis vector has a scalar or scalar function that creates some changing value depending on t. (5.1), for example, is the equation of a 2-D circle with radius r:

$$\mathbf{p}(t) = r\cos t\mathbf{i} + r\sin t\mathbf{j}, \quad t \in [0,\ 2\pi]. \tag{5.1}$$

With $r = 5$, (5.1) describes vectors for different values of t:

$$\mathbf{p}(0) = 5\mathbf{i}$$
$$\mathbf{p}(\pi/2) = 5\mathbf{j}$$

© Springer-Verlag London Ltd., part of Springer Nature 2021
J. Vince, *Vector Analysis for Computer Graphics*,
https://doi.org/10.1007/978-1-4471-7505-6_5

Fig. 5.1 Function
$\mathbf{p}(t) = 5\cos t\mathbf{i} + 5\sin t\mathbf{j}$,
with four vectors

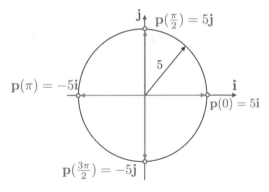

Fig. 5.2 Part of a helix
generated by $\mathbf{p}(t) =$
$5\cos t\mathbf{i} + 5\sin t\mathbf{j} + \frac{2t}{\pi}\mathbf{k}$

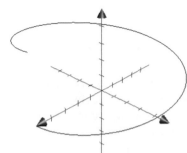

$$\mathbf{p}(\pi) = -5\mathbf{i}$$
$$\mathbf{p}(3\pi/2) = -5\mathbf{j}.$$

Figure 5.1 shows the curve generated by (5.1) with four vectors.

By adding a function for the **k** basis vector to (5.1), we create (5.2), which describes part of a helix:

$$\mathbf{p}(t) = r\cos t\mathbf{i} + r\sin t\mathbf{j} + \frac{ht}{2\pi}\mathbf{k}, \quad t \in [0,\ 2\pi]. \tag{5.2}$$

With $r = 5$, $h = 4$, (5.2) describes a helix, with the following position vectors for different values of t:

$$\mathbf{p}(0) = 5\mathbf{i} + 0\mathbf{j} + 0\mathbf{k}$$
$$\mathbf{p}(\pi/2) = 0\mathbf{i} + 5\mathbf{j} + 1\mathbf{k}$$
$$\mathbf{p}(\pi) = -5\mathbf{i} + 0\mathbf{j} + 2\mathbf{k}$$
$$\mathbf{p}(3\pi/2) = 0\mathbf{i} - 5\mathbf{j} + 3\mathbf{k}$$
$$\mathbf{p}(2\pi) = 5\mathbf{i} + 0\mathbf{j} + 4\mathbf{k}.$$

The curve is shown in Fig. 5.2.

5.3   Differentiating Vector-Valued Functions

When a vector-valued function's parameter represents time, it can reveal the velocity and acceleration of a point travelling along the function's curve. However, in order to reveal the velocity and acceleration, we need to differentiate a vector-valued function, which is what we do next.

We begin by defining the position of a point $P(x, y)$ on the plane using a vector:

$$\mathbf{p} = x\mathbf{i} + y\mathbf{j}$$

or a point $P(x, y, z)$ in 3-D space as

$$\mathbf{p} = x\mathbf{i} + y\mathbf{j} + z\mathbf{k}$$

and continue with 3-D vector-valued functions, but acknowledge that other spaces are equally valid.

If the point is controlled by a time-based function with parameter t, then the position vector for P has the form:

$$\mathbf{p}(t) = x(t)\mathbf{i} + y(t)\mathbf{j} + z(t)\mathbf{k}.$$

The derivative of $\mathbf{p}(t)$ is another vector formed from the derivatives of $x(t)$, $y(t)$ and $z(t)$:

$$\frac{d}{dt}\mathbf{p}(t) = \mathbf{p}'(t) = \frac{d}{dt}x(t)\mathbf{i} + \frac{d}{dt}y(t)\mathbf{j} + \frac{d}{dt}z(t)\mathbf{k}$$

which shows that the total derivative is the sum of the derivatives of the individual basis vectors.

For example, given

$$\mathbf{p}(t) = 10\sin t\mathbf{i} + 5t^2\mathbf{j} + 20\cos t\mathbf{k}$$

then,

$$\frac{d}{dt}\mathbf{p}(t) = 10\cos t\mathbf{i} + 10t\mathbf{j} - 20\sin t\mathbf{k}.$$

We are now in a position to reveal the velocity and acceleration from a vector-valued function.

5.3.1   Velocity and Speed

As the function $\mathbf{p}(t)$ locates a point at time t, its derivative is the rate of change of position with respect to time, i.e. its instantaneous velocity. For example, if $\mathbf{p}(t)$ is

Fig. 5.3 Instantaneous
velocity at time t

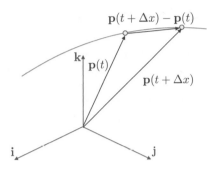

the position of a point P at time t, P's change in position from t to $t + \Delta t$ is

$$\Delta \mathbf{p} = \mathbf{p}(t + \Delta t) - \mathbf{p}(t).$$

Dividing by Δt:

$$\frac{\Delta \mathbf{p}}{\Delta t} = \frac{\mathbf{p}(t + \Delta t) - \mathbf{p}(t)}{\Delta t}.$$

In the limit as $\Delta t \to 0$ we have:

$$\frac{d}{dt}\mathbf{p}(t) = \mathbf{v}(t) = \lim_{\Delta t \to 0} \frac{\mathbf{p}(t + \Delta t) - \mathbf{p}(t)}{\Delta t}$$

which is the instantaneous velocity of P at time t. Figure 5.3 shows this diagrammatically.

For example, if the functions controlling a particle are $x(t) = 3 \cos t$, $y(t) = 5 \sin t$ and $z(t) = 2t$, then

$$\mathbf{p}(t) = 3 \cos t\mathbf{i} + 5 \sin t\mathbf{j} + 2t\mathbf{k}$$

and differentiating $\mathbf{p}(t)$ gives the instantaneous velocity vector:

$$\mathbf{v}(t) = \mathbf{p}'(t) = -3 \sin t\mathbf{i} + 5 \cos t\mathbf{j} + 2\mathbf{k}.$$

Figure 5.4 shows a point P moving along a trajectory defined by its position vector $\mathbf{p}(t)$. P's instantaneous velocity is represented by $\mathbf{v}(t)$ which is tangential to the trajectory at P.

If P's position is given by $\mathbf{p}(t)$ with the following derivatives:

$$\mathbf{p}(t) = x(t)\mathbf{i} + y(t)\mathbf{j} + z(t)\mathbf{k}$$
$$x'(t) = \frac{d}{dt}x(t)$$
$$y'(t) = \frac{d}{dt}y(t)$$

Fig. 5.4 Position and
instantaneous velocity of P
at time t

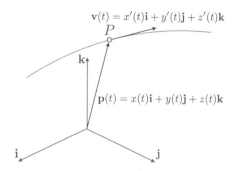

$$z'(t) = \frac{d}{dt}z(t)$$

the instantaneous speed of P is given by the magnitude of the instantaneous velocity
vector:

$$||\mathbf{v}(t)|| = \sqrt{(x'(t))^2 + (y'(t))^2 + (z'(t))^2}.$$

In the case of

$$\mathbf{p}(t) = 3\cos t\,\mathbf{i} + 5\sin t\,\mathbf{j} + 2t\,\mathbf{k}$$
$$\mathbf{v}(t) = -3\sin t\,\mathbf{i} + 5\cos t\,\mathbf{j} + 2\,\mathbf{k}$$

the instantaneous speed of P is:

$$||\mathbf{v}(t)|| = \sqrt{(-3\sin t)^2 + (5\cos t)^2 + 2^2}$$
$$= \sqrt{9\sin^2 t + 25\cos^2 t + 4}.$$

At time $t = 0$, the instantaneous speed of P is:

$$||\mathbf{v}(0)|| = \sqrt{25 + 4} = \sqrt{29}$$

and at time $t = \pi$, the instantaneous speed of P is:

$$||\mathbf{v}(\pi)|| = \sqrt{25 + 4} = \sqrt{29}.$$

The red lines in Fig. 5.5 represent the instantaneous velocity vectors at time $t = 0$
and $t = \pi$. At time $t = 0$, the coordinates of the line's base are given by:

$$\mathbf{p}(0) = 3\mathbf{i} + 0\mathbf{j} + 0\mathbf{k}, \quad (3,\ 0,\ 0)$$

the vector is:

$$\mathbf{v}(0) = 0\mathbf{i} + 5\mathbf{j} + 2\mathbf{k}$$

Fig. 5.5 Instantaneous
velocity vectors

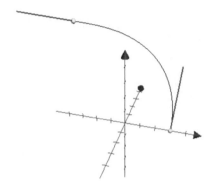

which makes the coordinates of the line's head: (3, 5, 2).

At time $t = \pi$, the coordinates of the line's base are given by:

$$\mathbf{p}(\pi) = -3\mathbf{i} + 0\mathbf{j} + 2\pi\mathbf{k}, \quad (-3, \ 0, \ 2\pi)$$

the vector is:

$$\mathbf{v}(\pi) = 0\mathbf{i} - 5\mathbf{j} + 2\mathbf{k}$$

which makes the coordinates of the line's head: $(-3, \ -5, \ 2 + 2\pi)$.

5.3.2   Acceleration

The instantaneous acceleration of a point with position vector $\mathbf{p}(t)$ is the second
derivative of $\mathbf{p}(t)$, or the derivative of the point's instantaneous velocity vector:

$$\mathbf{a}(t) = \mathbf{p}''(t) = \mathbf{v}'(t) = \frac{d^2}{dt^2}x(t)\mathbf{i} + \frac{d^2}{dt^2}y(t)\mathbf{j} + \frac{d^2}{dt^2}z(t)\mathbf{k}.$$

In the case of

$$\mathbf{p}(t) = 3\cos t\mathbf{i} + 5\sin t\mathbf{j} + 2t\mathbf{k}$$
$$\mathbf{v}(t) = -3\sin t\mathbf{i} + 5\cos t\mathbf{j} + 2\mathbf{k}$$
$$\mathbf{a}(t) = -3\cos t\mathbf{i} - 5\sin t\mathbf{j} + 0\mathbf{k}$$
$$||\mathbf{v}(t)|| = \sqrt{9\sin^2 t + 25\cos^2 t + 4}$$
$$||\mathbf{a}(t)|| = \sqrt{9\cos^2 t + 25\sin^2 t}.$$

At time $t = 0$:

$$\mathbf{p}(0) = 3\mathbf{i} + 0\mathbf{j} + 0\mathbf{k}$$
$$\mathbf{v}(0) = 0\mathbf{i} + 5\mathbf{j} + 2\mathbf{k}$$
$$\mathbf{a}(0) = -3\mathbf{i} + 0\mathbf{j} + 0\mathbf{k}$$
$$||\mathbf{v}(0)|| = \sqrt{25 + 4} = \sqrt{29}$$
$$||\mathbf{a}(0)|| = \sqrt{9} = 3.$$

At time $t = \pi/2$:

$$\mathbf{p}(\pi/2) = 0\mathbf{i} + 5\mathbf{j} + \pi\mathbf{k}$$
$$\mathbf{v}(\pi/2) = -3\mathbf{i} + 0\mathbf{j} + 2\mathbf{k}$$
$$\mathbf{a}(\pi/2) = 0\mathbf{i} - 5\mathbf{j} + 0\mathbf{k}$$
$$||\mathbf{v}(\pi/2)|| = \sqrt{9 + 4} = \sqrt{13}$$
$$||\mathbf{a}(\pi/2)|| = \sqrt{25} = 5.$$

At time $t = \pi$:

$$\mathbf{p}(\pi) = -3\mathbf{i} + 0\mathbf{j} + 2\pi\mathbf{k}$$
$$\mathbf{v}(\pi) = 0\mathbf{i} - 5\mathbf{j} + 2\mathbf{k}$$
$$\mathbf{a}(\pi) = 3\mathbf{i} + 0\mathbf{j} + 0\mathbf{k}$$
$$||\mathbf{v}(\pi)|| = \sqrt{25 + 4} = \sqrt{29}$$
$$||\mathbf{a}(\pi)|| = \sqrt{9} = 3.$$

Figure 5.6 shows the unit, instantaneous velocity vectors in red, and the unit, instantaneous acceleration vectors in green, for $t = 0$, $\pi/2$, π.

Let's examine one more time-based function before proceeding. For example, (5.3) describes the curve shown in Fig. 5.7:

Fig. 5.6 Instantaneous unit velocity (red) and unit acceleration (green) vectors

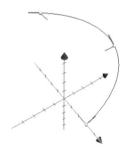

Fig. 5.7 Instantaneous
velocity (red) and
acceleration (green) unit
vectors

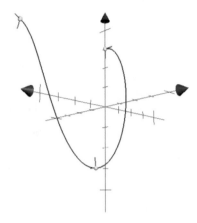

$$\mathbf{p}(t) = \tfrac{1}{3}t^3\mathbf{i} + 3\sin(2t)\mathbf{j} + 6\cos(2t)\mathbf{k}, \quad t \in [0,\ \pi]. \tag{5.3}$$

From (5.3) the instantaneous velocity and acceleration of a point are given by

$$\mathbf{v}(t) = t^2\mathbf{i} + 6\cos(2t)\mathbf{j} - 12\sin(2t)\mathbf{k}$$
$$\mathbf{a}(t) = 2t\mathbf{i} - 12\sin(2t)\mathbf{j} - 24\cos(2t)\mathbf{k}.$$

At time $t = 0$:

$$\mathbf{p}(0) = 0\mathbf{i} + 0\mathbf{j} + 6\mathbf{k}$$
$$\mathbf{v}(0) = 0\mathbf{i} + 6\mathbf{j} + 0\mathbf{k}$$
$$\mathbf{a}(0) = 0\mathbf{i} + 0\mathbf{j} - 24\mathbf{k}$$
$$||\mathbf{v}(0)|| = \sqrt{36} = 6$$
$$||\mathbf{a}(0)|| = \sqrt{24^2} = 24.$$

At time $t = \pi/2$:

$$\mathbf{p}(\pi/2) = \tfrac{\pi^3}{24}\mathbf{i} + 0\mathbf{j} - 6\mathbf{k}$$
$$\mathbf{v}(\pi/2) = \tfrac{\pi^2}{4}\mathbf{i} - 6\mathbf{j} + 0\mathbf{k}$$
$$\mathbf{a}(\pi/2) = \pi\mathbf{i} + 0\mathbf{j} + 24\mathbf{k}$$
$$||\mathbf{v}(\pi/2)|| = \sqrt{\tfrac{\pi^4}{16} + 36} \approx 6.488$$
$$||\mathbf{a}(\pi/2)|| = \sqrt{\pi^2 + 24^2} \approx 24.205.$$

At time $t = \pi$:

$$\mathbf{p}(\pi) = \tfrac{\pi^3}{3}\mathbf{i} + 0\mathbf{j} + 6\mathbf{k}$$

$$\mathbf{v}(\pi) = \pi^2\mathbf{i} + 6\mathbf{j} + 0\mathbf{k}$$
$$\mathbf{a}(\pi) = 2\pi\mathbf{i} + 0\mathbf{j} - 24\mathbf{k}$$
$$||\mathbf{v}(\pi)|| = \sqrt{\pi^4 + 36} \approx 11.55$$
$$||\mathbf{a}(\pi)|| = \sqrt{4\pi^2 + 24^2} \approx 24.809.$$

The three pairs of velocity and acceleration unit vectors are shown in Fig. 5.7.

5.3.3 Rules for Differentiating Vector-Valued Functions

Vector-valued functions are treated like vectors, in that they can be added, subtracted, scaled and multiplied, which leads to the following rules for their differentiation:

$$\frac{d}{dt}\left(\mathbf{p}(t) \pm \mathbf{q}(t)\right) = \frac{d}{dt}\mathbf{p}(t) \pm \frac{d}{dt}\mathbf{q}(t) \quad \text{addition and subtraction.}$$

$$\frac{d}{dt}\left(\lambda\mathbf{p}(t)\right) = \lambda\frac{d}{dt}\mathbf{p}(t) \quad \text{where } \lambda \in \mathbb{R}, \quad \text{scalar multiplier.}$$

$$\frac{d}{dt}\left(f(t)\mathbf{p}(t)\right) = f(t)\mathbf{p}'(t) + f'(t)\mathbf{p}(t) \quad \text{function multiplier.}$$

$$\frac{d}{dt}\left(\mathbf{p}(t) \cdot \mathbf{q}(t)\right) = \mathbf{p}(t) \cdot \mathbf{q}'(t) + \mathbf{p}'(t) \cdot \mathbf{q}(t) \quad \text{dot product.}$$

$$\frac{d}{dt}\left(\mathbf{p}(t) \times \mathbf{q}(t)\right) = \mathbf{p}(t) \times \mathbf{q}'(t) + \mathbf{p}'(t) \times \mathbf{q}(t) \quad \text{cross product.}$$

$$\frac{d}{dt}\left(\mathbf{p}(f(t))\right) = \mathbf{p}'(f(t))f'(t) \quad \text{function of a function.}$$

5.4 Integrating Vector-Valued Functions

The integral of a vector-valued function is its antiderivative, where each term is integrated individually. For example, given

$$\mathbf{p}(t) = x(t)\mathbf{i} + y(t)\mathbf{i} + z(t)\mathbf{k}$$

then,

$$\int_a^b \mathbf{p}(t)\,dt = \int_a^b x(t)\mathbf{i}\,dt + \int_a^b y(t)\mathbf{i}\,dt + \int_a^b z(t)\mathbf{k}\,dt.$$

Similarly,

$$\int \mathbf{p}(t)\,dt = \int x(t)\mathbf{i}\,dt + \int y(t)\mathbf{i}\,dt + \int z(t)\mathbf{k}\,dt + \mathbf{C}.$$

Integrating the velocity vector used above:

$$\mathbf{v}(t) = -3\sin t\mathbf{i} + 5\cos t\mathbf{j} + 2\mathbf{k}$$

then,

$$\int \mathbf{v}(t)\, dt = \int -3\sin t\mathbf{i}\, dt + \int 5\cos t\mathbf{j}\, dt + \int 2\mathbf{k}\, dt + \mathbf{C}$$

$$= -3\int \sin t\mathbf{i}\, dt + 5\int \cos t\mathbf{j}\, dt + 2\int 1\mathbf{k}\, dt + \mathbf{C}$$

$$= 3\cos t\mathbf{i} + 5\sin t\mathbf{j} + 2t\mathbf{k} + \mathbf{C}.$$

And if we know that $\mathbf{p}(0)$ points to $(3, \ 0, \ 0)$, then $\mathbf{p}(t) = 3\cos t\mathbf{i} + 5\sin t\mathbf{j} + 2t\mathbf{k}$. We have already seen that

$$\mathbf{v}(t) = \frac{d}{dt}\mathbf{p}(t)$$

$$\mathbf{a}(t) = \frac{d}{dt}\mathbf{v}(t)$$

therefore,

$$\mathbf{p}(t) = \int \mathbf{v}(t)\, dt$$

$$\mathbf{v}(t) = \int \mathbf{a}(t)\, dt.$$

5.4.1   Velocity of a Falling Object

When an object falls under the influence of gravity (9.8 m/s^2) for 3 seconds, its velocity at time t is given by

$$\mathbf{v}(t) = \int 9.8\, dt = 9.8t + C_1.$$

Assuming that its initial velocity is zero, then $\mathbf{v}(0) = 0$, and $C_1 = 0$. Therefore,

$$\mathbf{p}(t) = \int 9.8t\, dt = \tfrac{9.8}{2}t^2 + C_2 = 4.9t^2 + C_2.$$

But $\mathbf{p}(0) = 0$, and $C_2 = 0$, therefore,

$$\mathbf{p}(t) = 4.9t^2.$$

Consequently, after 3 seconds, the object has fallen $4.9 \times 3^2 = 40.1$m.

If the object had been given an initial downward velocity of 1 m/s, then $C_1 = 1$, which means that

$$\mathbf{p}(t) = \int 9.8t + 1 \, dt = \frac{9.8}{2}t^2 + t + C_2 = 4.9t^2 + t + C_2.$$

But $\mathbf{p}(0) = 0$, and $C_2 = 0$, therefore,

$$\mathbf{p}(t) = 4.9t^2 + t.$$

Consequently, after 3 seconds, the object has fallen $4.9 \times 3^2 + 3 = 43.1$ m.

5.4.2 Position of a Moving Object

Let's compute an object's position after 2 seconds if it is following a parametric curve such that its velocity is:

$$\mathbf{v}(t) = t^2\mathbf{i} + t\mathbf{j} + t^3\mathbf{k}$$

starting at the origin at time $t = 0$:

$$\mathbf{p}(t) = \int \mathbf{v}(t) \, dt + \mathbf{C}$$
$$= \int t^2\mathbf{i} + t\mathbf{j} + t^3\mathbf{k} \, dt + \mathbf{C}$$
$$= \int t^2\mathbf{i} \, dt + \int t\mathbf{j} \, dt + \int t^3\mathbf{k} \, dt + \mathbf{C}$$
$$= \frac{1}{3}t^3\mathbf{i} + \frac{1}{2}t^2\mathbf{j} + \frac{1}{4}t^4\mathbf{k} + \mathbf{C}.$$

But $\mathbf{p}(0) = 0\mathbf{i} + 0\mathbf{j} + 0\mathbf{k}$, therefore, the vector $\mathbf{C} = 0\mathbf{i} + 0\mathbf{j} + 0\mathbf{k}$, and

$$\mathbf{p}(t) = \frac{1}{3}t^3\mathbf{i} + \frac{1}{2}t^2\mathbf{j} + \frac{1}{4}t^4\mathbf{k}.$$

Consequently, after 2 seconds, the object is at

$$\mathbf{p}(2) = \frac{1}{3}2^3\mathbf{i} + \frac{1}{2}2^2\mathbf{j} + \frac{1}{4}2^4\mathbf{k}$$
$$= \frac{8}{3}\mathbf{i} + 2\mathbf{j} + 4\mathbf{k}.$$

Which is the point $\left(\frac{8}{3}, \ 2, \ 4\right)$.

5.5 Summary

The calculus of vector-valued functions is a large and complex subject, and in this short chapter we have only covered the basic principles for differentiating and integrating simple functions, which are summarised next.

5.5.1 Summary of Formulae

Given a function of the form:

$$\mathbf{p}(t) = x(t)\mathbf{i} + y(t)\mathbf{j} + z(t)\mathbf{k}$$

its derivative is:

$$\frac{d}{dt}\mathbf{p}(t) = \mathbf{p}'(t) = x'(t)\mathbf{i} + y'(t)\mathbf{j} + z'(t)\mathbf{k}$$

its integral is:

$$\int \mathbf{p}(t)\,dt = \int x(t)\mathbf{i}\,dt + \int y(t)\mathbf{i}\,dt + \int z(t)\mathbf{k}\,dt + \mathbf{C}$$

and definite integral:

$$\int_a^b \mathbf{p}(t)\,dt = \int_a^b x(t)\mathbf{i}\,dt + \int_a^b y(t)\mathbf{i}\,dt + \int_a^b z(t)\mathbf{k}\,dt.$$

If $\mathbf{p}(t)$ is a time-based position vector, its derivative is a velocity vector, and its second derivative is an acceleration vector:

$$\mathbf{p}(t) = x(t)\mathbf{i} + y(t)\mathbf{j} + z(t)\mathbf{k}$$
$$\mathbf{v}(t) = x'(t)\mathbf{i} + y'(t)\mathbf{j} + z'(t)\mathbf{k}$$
$$\mathbf{a}(t) = x''(t)\mathbf{i} + y''(t)\mathbf{j} + z''(t)\mathbf{k}.$$

The magnitude of $\mathbf{v}(t)$ represents the instantaneous speed:

$$\|\mathbf{v}(t)\| = \sqrt{(x'(t))^2 + (y'(t))^2 + (z'(t))^2}$$

and for the instantaneous acceleration:

$$\|\mathbf{a}(t)\| = \sqrt{(x''(t))^2 + (y''(t))^2 + (z''(t))^2}.$$

References

1. Vince J (2020) Foundation mathematics for computer science, 2nd edn. Springer. ISBN 978-3-030-42077-2
2. Vince J (2017) Mathematics for computer graphics, 5th edn. Springer. ISBN 978-1-4471-7334-2
3. Vince J (2019) Calculus for computer graphics, 2nd edn. Springer. ISBN 978-3-030-11375-9

Chapter 6
Vector Differential Operators

6.1 Introduction

This chapter covers the mathematical objects: scalar and vector fields, and the three vector differential operators: grad, div and curl. It begins with an overview of scalar and vector fields, followed by the gradient of a scalar field, the divergence and curl of a vector field. The descriptions assume the reader is familiar with calculus, but those wishing to brush up their knowledge will find the following books useful [1], [2], [3].

The figures employ a simple colour key where blue equates with a low scalar value or small vector, moving through the rainbow to red, which equates with a high scalar value or large vector.

6.2 Scalar Fields

A *scalar field* is a set of scalars normally associated with physical space, where such scalars are derived empirically, or from mathematical functions. Some examples of empirical scalar fields include:

- Bank of England interest rate over time.
- Barometric pressure or temperature in a weather map.
- Depth of oceans over the surface of the Earth.
- Height, weight, BMI, age, gender, etc., for a group of people.
  The data for the above examples will be in various formats such as one-, two-, three-dimensional arrays, Cartesian coordinates, spherical coordinates, etc.

Figure 6.1 shows an image of a scalar field in $\mathbb{R}^2$ described by (6.1):

$$f(x, y) = \sin(xy) \qquad (6.1)$$

© Springer-Verlag London Ltd., part of Springer Nature 2021
J. Vince, *Vector Analysis for Computer Graphics*,
https://doi.org/10.1007/978-1-4471-7505-6_6

Fig. 6.1 Scalar field in $\mathbb{R}^2$
for $f(x, y) = \sin(xy)$

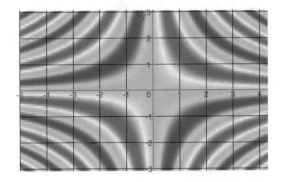

Fig. 6.2 Scalar field in $\mathbb{R}^3$
for $f(x, y, z) = \sin(xyz)$

where,

$$f(1, 1) = \sin(1) \approx 0.8415, \quad \text{is an orange-red pixel,}$$
$$f(-1, 1) = \sin(-1) \approx -0.8415, \quad \text{is a light-blue pixel.}$$

Figure 6.2 shows an image of a scalar field in $\mathbb{R}^3$ described by (6.2):

$$f(x, y, z) = \sin(xyz) \tag{6.2}$$

where,

$$f(1, 1, 1) = \sin(1) \approx 0.8415, \quad \text{is the nearest orange-red voxel,}$$
$$f(-1, -1, 1) = \sin(1) \approx 0.8415, \quad \text{is the farthest orange-red voxel.}$$

Although Figs. 6.1 and 6.2 show granularity, the mathematical functions are continuous over the fields they describe.

6.3   Vector Fields

A *vector field* is a set of vectors normally associated with physical space, where such vectors are derived empirically, or from mathematical functions. Some examples of empirical vector fields include:

- Speed and direction of people in a photograph.
- Motion of gases in a 3-D volume.
- Wind magnitude and direction over the surface of the Earth.
- Intensity and direction of a magnetic field around an electrical conductor.

Figure 6.3 shows a vector field in $\mathbb{R}^2$ described by (6.3):

$$\mathbf{F}(x, y) = \sin y \mathbf{i} + \sin x \mathbf{j} \tag{6.3}$$

where,

$$\mathbf{F}(1, 1) = \sin(1)\mathbf{i} + \sin(1)\mathbf{j} = \begin{bmatrix} 0.8415 \\ 0.8415 \end{bmatrix}, \quad \text{is shown as a blue arrow.}$$

$$\mathbf{F}(-1, 1) = \sin(1)\mathbf{i} + \sin(-1)\mathbf{j} = \begin{bmatrix} 0.8415 \\ -0.8415 \end{bmatrix}, \quad \text{is shown as a green arrow.}$$

Figure 6.4 shows a vector field in $\mathbb{R}^3$ described by (6.4):

$$\mathbf{F}(x, y, z) = x\mathbf{i} + y\mathbf{j} + z\mathbf{k} \tag{6.4}$$

where,

$$\mathbf{F}(5, 5, 5) = 5\mathbf{i} + 5\mathbf{j} + 5\mathbf{k}, \quad \text{is shown red.}$$
$$\mathbf{F}(5, -5, 5) = 5\mathbf{i} - 5\mathbf{j} + 5\mathbf{k}, \quad \text{is shown blue.}$$

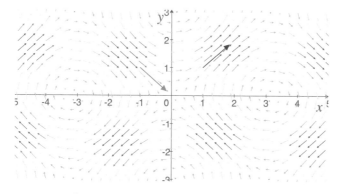

Fig. 6.3   Vector field in $\mathbb{R}^2$ for $\mathbf{F}(x, y) = \sin y \mathbf{i} + \sin x \mathbf{j}$

Fig. 6.4 Vector field in $\mathbb{R}^3$ for $\mathbf{F}(x, y, z) = x\mathbf{i} + y\mathbf{j} + z\mathbf{k}$

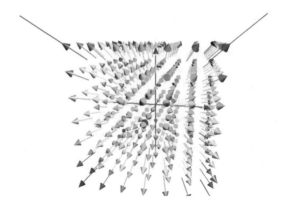

3-D scalar and vector fields are difficult to illustrate using a single image, and are best seen using a computer capable of animating the images in real time.

Chapter 5 shows how vector-valued functions are differentiated and integrated, and now we introduce three differential vector operators: grad, div and curl, which reveal features of scalar and vector fields.

6.4   The Gradient of a Scalar Field

The *grad* or gradient differential vector operator computes the gradient of a scalar field, and is represented by the ∇ symbol, and called either 'del' or 'nabla'. When we are dealing with functions of the form $f(x, y)$ or $f(x, y, z)$, we assume that f is differentiable at all points (x, y) or (x, y, z). For example, given a scalar field in $\mathbb{R}^2$ described by $f(x, y)$, each (x, y) creates a scalar value, and $\nabla f(x, y)$ shows the scalar gradient at this point in the form of a 2-D vector. Similarly, for a scalar field in $\mathbb{R}^3$ described by $f(x, y, z)$, each (x, y, z) creates a scalar value, and $\nabla f(x, y, z)$ shows the scalar gradient at this point in the form of a 3-D vector.

In one, two and three dimensions, ∇ is defined as:

$$\nabla = \left[\frac{\partial}{\partial x} \right] = \mathbf{i}\frac{\partial}{\partial x} \equiv \frac{\partial}{\partial x}\mathbf{i}$$

$$\nabla = \begin{bmatrix} \dfrac{\partial}{\partial x} \\ \dfrac{\partial}{\partial y} \end{bmatrix} = \mathbf{i}\frac{\partial}{\partial x} + \mathbf{j}\frac{\partial}{\partial y} \equiv \frac{\partial}{\partial x}\mathbf{i} + \frac{\partial}{\partial y}\mathbf{j}$$

$$\nabla = \begin{bmatrix} \dfrac{\partial}{\partial x} \\ \dfrac{\partial}{\partial y} \\ \dfrac{\partial}{\partial z} \end{bmatrix} = \mathbf{i}\frac{\partial}{\partial x} + \mathbf{j}\frac{\partial}{\partial y} + \mathbf{k}\frac{\partial}{\partial z} \equiv \frac{\partial}{\partial x}\mathbf{i} + \frac{\partial}{\partial y}\mathbf{j} + \frac{\partial}{\partial z}\mathbf{k}.$$

Fig. 6.5 Scalar field in $\mathbb{R}^2$
for the function
$f(x, y) = 0.8x^2 y$

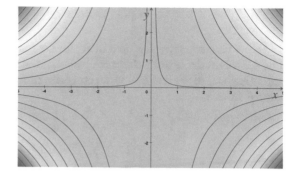

Fig. 6.6 Vector field in $\mathbb{R}^2$
created by
$\nabla f(x, y) = 1.6xy\mathbf{i} + 0.8x^2\mathbf{j}$

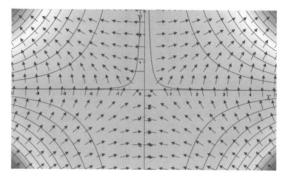

If the function $f(x, y, z)$ represents a scalar field, then ∇f creates a vector field for different values of (x, y, z):

$$\nabla f(x, y, z) = \frac{\partial f}{\partial x}\mathbf{i} + \frac{\partial f}{\partial y}\mathbf{j} + \frac{\partial f}{\partial z}\mathbf{k}.$$

6.4.1 *Gradient of a Scalar Field in* $\mathbb{R}^2$

Let's illustrate the gradient of a scalar field in $\mathbb{R}^2$ with two examples.

Figure 6.5, shows a scalar field for the function $f(x, y) = 0.8x^2 y$, and Fig. 6.6 shows the 2-D vector field created by $\nabla f(x, y)$, calculated as follows:

$$f(x, y) = 0.8x^2 y$$
$$\nabla f(x, y) = \frac{\partial f}{\partial x}\mathbf{i} + \frac{\partial f}{\partial y}\mathbf{j}$$
$$\frac{\partial f}{\partial x} = 1.6xy$$
$$\frac{\partial f}{\partial y} = 0.8x^2$$

$$\nabla f(x, y) = 1.6xy\mathbf{i} + 0.8x^2\mathbf{j}.$$

Figure 6.7 shows a scalar field in $\mathbb{R}^2$ for the function $f(x, y) = \sqrt{x^2 + y^2}$, and Fig. 6.8 shows the vector field in $\mathbb{R}^2$, calculated as follows. Notice that the vectors are orthogonal to the constant value contours; this is exploited to compute surface normal vectors.

We begin by letting $u = x^2 + y^2$, then $f(x, y) = u^{\frac{1}{2}}$, and differentiating:

$$\frac{\partial f}{\partial x} = \frac{\partial f}{\partial u}\frac{\partial u}{\partial x}$$

$$\frac{\partial f}{\partial u} = \frac{1}{2\sqrt{u}} = \frac{1}{2\sqrt{x^2 + y^2}}$$

$$\frac{\partial u}{\partial x} = 2x$$

$$\frac{\partial f}{\partial x} = \frac{x}{\sqrt{x^2 + y^2}}$$

$$\frac{\partial f}{\partial y} = \frac{y}{\sqrt{x^2 + y^2}}$$

Fig. 6.7 Scalar field in $\mathbb{R}^2$ for the function $f(x, y) = \sqrt{x^2 + y^2}$

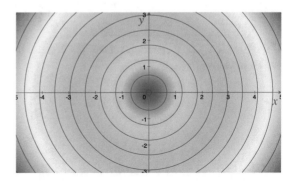

Fig. 6.8 Vector field in $\mathbb{R}^2$ created by $\nabla f(x, y) = \frac{x}{\sqrt{x^2+y^2}}\mathbf{i} + \frac{y}{\sqrt{x^2+y^2}}\mathbf{j}$

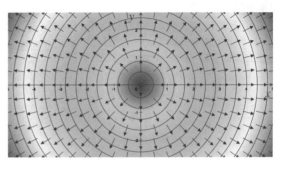

$$\nabla f(x, y) = \frac{\partial f}{\partial x}\mathbf{i} + \frac{\partial f}{\partial y}\mathbf{j}$$

$$= \frac{x}{\sqrt{x^2 + y^2}}\mathbf{i} + \frac{y}{\sqrt{x^2 + y^2}}\mathbf{j}.$$

6.4.2 Gradient of a Scalar Field in $\mathbb{R}^3$

Figure 6.9 shows a simple scalar field in $\mathbb{R}^3$ for the function function $f(x, y, z) = xy + yz$, and Fig. 6.10 shows the resulting vector field, calculated as follows:

$$f(x, y, z) = xy + yz$$

$$\nabla f(x, y) = \frac{\partial f}{\partial x}\mathbf{i} + \frac{\partial f}{\partial y}\mathbf{j} + \frac{\partial f}{\partial z}\mathbf{k}$$

$$\frac{\partial f}{\partial x} = y$$

Fig. 6.9 Scalar field in $\mathbb{R}^3$
for the function
$f(x, y, z) = xy + yz$

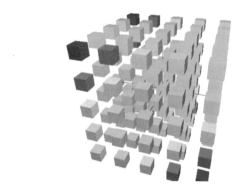

Fig. 6.10 Vector field in $\mathbb{R}^3$
for the function $f(x, y, z) = y\mathbf{i} + (x + z)\mathbf{j} + y\mathbf{k}$

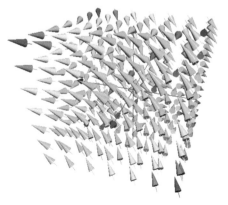

$$\frac{\partial f}{\partial y} = x + z$$

$$\frac{\partial f}{\partial z} = y$$

$$\nabla f(x, y) = y\mathbf{i} + (x + z)\mathbf{j} + y\mathbf{k}.$$

6.4.3   Surface Normal Vectors

One useful application of the grad differential operator is creating a vector orthogonal to a curve in $\mathbb{R}^2$, or a surface in $\mathbb{R}^3$. If a function is of the form $f(x, y, z) = $ constant, for example the equation of a circle: $f(x, y) = x^2 + y^2 = r^2$, or the equation of a sphere: $f(x, y, z) = x^2 + y^2 + z^2 = r^2$, ∇f creates a vector normal to the function at any point.

Figure 6.11 shows an array of scalars for the function $f(x, y) = x^2 + y^2$. For example, the cell with coordinates $(5, 5)$ contains 50, because $50 = 5^2 + 5^2$. In reality, a scalar field is continuous, rather than discrete, as shown in Fig. 6.11.

One can see from Fig. 6.11 that this scalar field comprises a family of concentric contours, one, of which, is sketched in the figure.

Taking the partial derivative of $f(x, y) = x^2 + y^2$ in the x-direction:

$$\frac{\partial f}{\partial x} = 2x$$

gives the instantaneous rate of change at any point (x, y) irrespective of the value of y. Similarly, taking the partial derivative of $f(x, y) = x^2 + y^2$ in the y-direction:

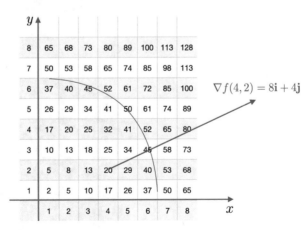

Fig. 6.11   A scalar field for $f(x, y) = x^2 + y^2$

$$\frac{\partial f}{\partial y} = 2y$$

gives the instantaneous rate of change at any point (x, y) irrespective of the value of x.

But we know that the differential operator ∇ maps a scalar field to a vector field using these partial derivatives as follows:

$$\nabla f(x, y) = \begin{bmatrix} \dfrac{\partial f}{\partial x} \\ \dfrac{\partial f}{\partial y} \end{bmatrix} = \frac{\partial f}{\partial x}\mathbf{i} + \frac{\partial f}{\partial y}\mathbf{j}$$

$$= 2x\mathbf{i} + 2y\mathbf{j}.$$

For example, when $x = 4$ and $y = 2$:

$$\nabla f(4, 2) = 8\mathbf{i} + 4\mathbf{j}$$

which is sketched in Fig. 6.11. Observe that the vector is orthogonal to the contour. In fact, all vectors are orthogonal to all such contours. In other words, the vector is normal to any curve defined by the function $f(x, y) = x^2 + y^2$. For example, the equation of a circle is

$$x^2 + y^2 = r^2$$

where r is the radius. Therefore, we can create a function

$$f(x, y) = x^2 + y^2 - r^2 = 0.$$

Therefore,

$$\nabla f = 2x\mathbf{i} + 2y\mathbf{j}$$

which is the normal vector at (x, y), as shown in Fig. 6.11.

Although the above reasoning seems acceptable, here is the mathematical proof. Given a surface $f(x, y, z) = c$, where c is a constant, let $P(x, y, z)$ be a point on the surface with position vector $\mathbf{p} = x\mathbf{i} + y\mathbf{j} + z\mathbf{k}$. Then

$$\nabla f = \frac{\partial f}{\partial x}\mathbf{i} + \frac{\partial f}{\partial y}\mathbf{j} + \frac{\partial f}{\partial z}\mathbf{k}$$

$$d\mathbf{p} = dx\mathbf{i} + dy\mathbf{j} + dz\mathbf{k}$$

and $d\mathbf{p}$ must lie in the tangent plane at P. But

$$df = \frac{\partial f}{\partial x}dx + \frac{\partial f}{\partial y}dy + \frac{\partial f}{\partial z}dz = 0$$

which can be expressed as the dot product:

$$\left(\frac{\partial f}{\partial x}\mathbf{i} + \frac{\partial f}{\partial y}\mathbf{j} + \frac{\partial f}{\partial z}\mathbf{k}\right) \cdot (dx\mathbf{i} + dy\mathbf{j} + dz\mathbf{k}) = 0$$

$$\nabla f \cdot d\mathbf{p} = 0.$$

For $\nabla f \cdot d\mathbf{p} = 0$ to be true, they must be orthogonal.

In general, given a function $f(x, y) = c$ or $f(x, y, z) = c$, then a normal vector $\mathbf{n}$ is

$$\mathbf{n} = \nabla f$$

and a unit normal vector is

$$\hat{\mathbf{n}} = \frac{\nabla f}{\|\nabla f\|}.$$

For 3-D functions, such as $f(x, y, z) = 2xy + 3z = 0$:

$$\frac{\partial f}{\partial x} = 2y$$

$$\frac{\partial f}{\partial y} = 2x$$

$$\frac{\partial f}{\partial z} = 3$$

$$\nabla f = 2y\mathbf{i} + 2x\mathbf{j} + 3\mathbf{k}.$$

Figure 6.12 shows the surface of the function $f(x, y, z) = 2xy + 3z = 0$, and two normal vectors calculated as follows:

Fig. 6.12 The surface of the function
$f(x, y, z) = 2xy + 3z = 0$

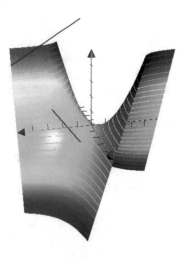

When $x = y = 1$, then $z = -2/3$, then $\nabla f = 2\mathbf{i} + 2\mathbf{j} + 3\mathbf{k}$, which is shown in red.

When $x = 5$ and $z = 4$, then $y = -1.2$, then $\nabla f = -2.4\mathbf{i} + 10\mathbf{j} + 3\mathbf{k}$, which is shown in blue.

6.5 The Divergence of a Vector Field

Illustrations of vector fields are only a crude approximation of what is actually happening in space. To begin with, vectors in the form of arrows are an artifice, as they attempt to simulate the magnitude and direction of something represented by a vector-valued function, be it a liquid, dust storm, electromagnetic force, etc. A liquid such as water, is not compressible under normal conditions, so if we inspect a specific volume of water, the number of water molecules entering the volume is balanced by the number of water molecules leaving. In this case, we say the divergence of the vector field at any point is zero. See Fig. 6.13.

In the case of airborne dust particles, they are compressible, as the number of particles in a unit volume of air depends on the local pressure forces, so if we inspect a specific volume of dust, the number of dust particles entering the volume, is different to the number of dust particles leaving. In this case, we say the divergence of the vector field changes from point to point. See Fig. 6.14.

Fig. 6.13 A 2-D vector field simulating water movement

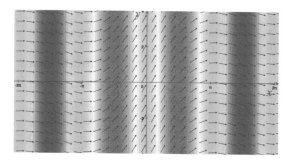

Fig. 6.14 A vector field simulating a dust storm

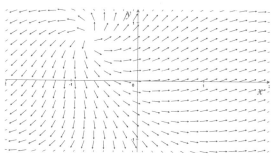

Consider a 3-D vector field $\mathbf{F}(x, y, z)$ defined by three functions $R(x, y, z)$, $S(x, y, z)$, $T(x, y, z)$, (6.5):

$$\mathbf{F} = R\mathbf{i} + S\mathbf{j} + T\mathbf{k}. \tag{6.5}$$

The divergence of $\mathbf{F}$ is defined as:

$$\text{div } \mathbf{F} = \frac{\partial R}{\partial x}\mathbf{i} + \frac{\partial S}{\partial y}\mathbf{j} + \frac{\partial T}{\partial z}\mathbf{k}. \tag{6.6}$$

Equation (6.6) can also be written as a scalar (dot) product:

$$\text{div } \mathbf{F} = \left(\frac{\partial}{\partial x}\mathbf{i} + \frac{\partial}{\partial y}\mathbf{j} + \frac{\partial}{\partial z}\mathbf{k}\right) \cdot \left(R\mathbf{i} + S\mathbf{j} + T\mathbf{k}\right) \tag{6.7}$$

and substituting the differential operator ∇, (6.7) becomes:

$$\text{div } \mathbf{F} = \nabla \cdot \mathbf{F}.$$

For example, let's find the divergence of the vector field $\mathbf{F} = 3x\mathbf{i} - xy\mathbf{j} + 2yz^2\mathbf{k}$:

$$\nabla \cdot \mathbf{F} = \frac{\partial}{\partial x}(3x) + \frac{\partial}{\partial y}(-xy) + \frac{\partial}{\partial z}\left(2yz^2\right)$$
$$= 3 - x + 4yz.$$

Substituting different values of x, y, z, we get

$$\nabla \cdot \mathbf{F}(0, 0, 0) = 3$$
$$\nabla \cdot \mathbf{F}(4, 0, 0) = -1$$
$$\nabla \cdot \mathbf{F}(-4, 0, 0) = 7$$
$$\nabla \cdot \mathbf{F}(4, 4, 0) = -1$$
$$\nabla \cdot \mathbf{F}(4, -4, 0) = -1$$
$$\nabla \cdot \mathbf{F}(4, 4, 4) = 63$$

which shows that the divergence of $\mathbf{F}$ changes from point to point. Figure 6.15 shows the divergence values for $\nabla \cdot \mathbf{F} = 3 - x + 4yz$, with the top red cube equal to $(5, 5, 5)$.

Fig. 6.15 The scalar field
for $\nabla \cdot \mathbf{F} = 3 - x + 4yz$

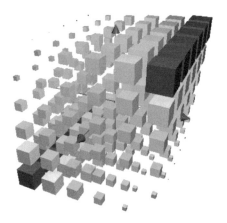

Fig. 6.16 The vector field
for
$\mathbf{F}(x, y, z) = x\mathbf{i} + y\mathbf{j} + z\mathbf{k}$

Now let's consider the function $\mathbf{F}(x, y, z) = x\mathbf{i} + y\mathbf{j} + z\mathbf{k}$, shown in Fig. 6.16, where

$$\nabla \cdot \mathbf{F} = \frac{\partial}{\partial x}x + \frac{\partial}{\partial y}y + \frac{\partial}{\partial z}z$$
$$= 1 + 1 + 1$$
$$= 3.$$

Which shows that the divergence is positive when the vector flow is away from a point. Similarly, the divergence is negative when the vector flow is towards a point.

6.6   Curl of a Vector Field

Having seen that the differential operator ∇ can be used with a scalar field ∇F, and in conjunction with a vector field as a dot product $\nabla \cdot \mathbf{F}$, it is only natural to wonder if it can be used with a vector field with the cross product. Well it can, and creates what is known as the *curl* of a vector field, and is defined as

$$\text{curl } \mathbf{F} = \nabla \times \mathbf{F}.$$

If $\mathbf{F} = R\mathbf{i} + S\mathbf{j} + T\mathbf{k}$, and using a determinant to represent a cross product, we have:

$$\nabla \times \mathbf{F} = \begin{vmatrix} \mathbf{i} & \mathbf{j} & \mathbf{k} \\ \dfrac{\partial}{\partial x} & \dfrac{\partial}{\partial y} & \dfrac{\partial}{\partial z} \\ R & S & T \end{vmatrix}$$

$$= \left(\frac{\partial T}{\partial y} - \frac{\partial S}{\partial z} \right)\mathbf{i} - \left(\frac{\partial T}{\partial x} - \frac{\partial R}{\partial z} \right)\mathbf{j} + \left(\frac{\partial S}{\partial x} - \frac{\partial R}{\partial y} \right)\mathbf{k}.$$

For example, with $\mathbf{F} = -y\mathbf{i} + x\mathbf{j} + 0\mathbf{k}$, then $R = -y\mathbf{i}$, $S = x\mathbf{j}$ and $T = 0\mathbf{k}$, then

$$\frac{\partial T}{\partial y} = 0, \quad \frac{\partial S}{\partial z} = 0$$

$$\left(\frac{\partial T}{\partial y} - \frac{\partial S}{\partial z} \right)\mathbf{i} = 0\mathbf{i}$$

$$\frac{\partial T}{\partial x} = 0, \quad \frac{\partial R}{\partial z} = 0$$

$$-\left(\frac{\partial T}{\partial x} - \frac{\partial R}{\partial z} \right)\mathbf{j} = 0\mathbf{j}$$

$$\frac{\partial S}{\partial x} = 1, \quad \frac{\partial R}{\partial y} = -1$$

$$\left(\frac{\partial S}{\partial x} - \frac{\partial R}{\partial y} \right)\mathbf{k} = 2\mathbf{k}$$

$$\nabla \times \mathbf{F} = 0\mathbf{i} + 0\mathbf{j} + 2\mathbf{k}.$$

Figure 6.17 shows the vector field for $\mathbf{F} = -y\mathbf{i} + x\mathbf{j} + 0\mathbf{k}$ looking down the $\mathbf{k}$ axis. The image also includes several unit vectors drawn in mauve using the vector function, which confirm that the vector function possesses rotational qualities. It is this *curling* of the vector field about a vector - in this case the $\mathbf{k}$-axis - that $\nabla \times \mathbf{F}$ measures. A positive value indicates a counter-clockwise rotation, and a negative value, a clockwise rotation, and a zero value indicates no rotation.

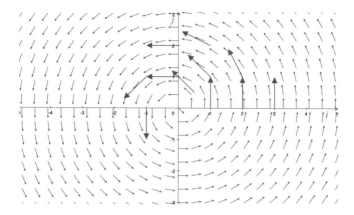

Fig. 6.17 The vector field for $\mathbf{f} = -y\mathbf{i} + x\mathbf{j} + 0\mathbf{k}$

Fig. 6.18 The vector field
for $\mathbf{F} = -y\mathbf{i} + x\mathbf{j} + 0\mathbf{k}$

Figure 6.18 shows a 3-D view of the same vector function, looking down the **k**-axis. Once again, it is clear that the vector field is rotating about the **k**-axis, and the curl $\nabla \times \mathbf{F} = 2\mathbf{k}$ is a measure of this rotation.

Now let's consider a similar vector field $\mathbf{F} = 0\mathbf{i} + z\mathbf{j} - y\mathbf{k}$, then $R = 0\mathbf{i}$, $S = z\mathbf{j}$ and $T = -y\mathbf{k}$, that is rotating about the **i**-axis, then:

$$\frac{\partial T}{\partial y} = -1, \quad \frac{\partial S}{\partial z} = 1$$

$$\left(\frac{\partial T}{\partial y} - \frac{\partial S}{\partial z} \right)\mathbf{i} = -2\mathbf{i}$$

$$\frac{\partial T}{\partial x} = 0, \quad \frac{\partial R}{\partial z} = 0$$

$$-\left(\frac{\partial T}{\partial x} - \frac{\partial R}{\partial z} \right)\mathbf{j} = 0\mathbf{j}$$

Fig. 6.19 The vector field
for $\mathbf{F} = 0\mathbf{i} + z\mathbf{j} - y\mathbf{k}$

$$\frac{\partial S}{\partial x} = 0, \quad \frac{\partial R}{\partial y} = 0$$

$$\left(\frac{\partial S}{\partial x} - \frac{\partial R}{\partial y}\right)\mathbf{k} = 0\mathbf{k}$$

$$\nabla \times \mathbf{F} = -2\mathbf{i} + 0\mathbf{j} + 0\mathbf{k}.$$

This time the result is negative, which is shown in Fig. 6.19 looking down the $\mathbf{i}$-axis, as a clockwise rotation.

Finally, if the vector function is $\mathbf{F} = x\mathbf{i} + y\mathbf{j} + 0\mathbf{k}$, then $R = x\mathbf{i}$, $S = y\mathbf{j}$ and $T = 0\mathbf{k}$, then:

$$\frac{\partial T}{\partial y} = 0, \quad \frac{\partial S}{\partial z} = 0$$

$$\left(\frac{\partial T}{\partial y} - \frac{\partial S}{\partial z}\right)\mathbf{i} = 0\mathbf{i}$$

$$\frac{\partial T}{\partial x} = 0, \quad \frac{\partial R}{\partial z} = 0$$

$$-\left(\frac{\partial T}{\partial x} - \frac{\partial R}{\partial z}\right)\mathbf{j} = 0\mathbf{j}$$

$$\frac{\partial S}{\partial x} = 0, \quad \frac{\partial R}{\partial y} = 0$$

$$\left(\frac{\partial S}{\partial x} - \frac{\partial R}{\partial y}\right)\mathbf{k} = 0\mathbf{k}$$

$$\nabla \times \mathbf{F} = 0\mathbf{i} + 0\mathbf{j} + 0\mathbf{k}.$$

The result is zero curl, as confirmed by Fig. 6.20.

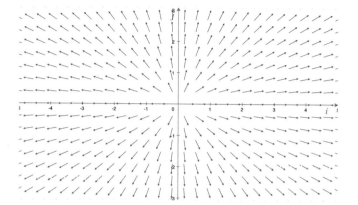

Fig. 6.20 The vector field for $\mathbf{F} = x\mathbf{i} + y\mathbf{j} + 0\mathbf{k}$

6.7 Worked Examples

6.7.1 Gradient of a Scalar Field

Calculate the gradient of the scalar field $f(x, y, z) = x^2 + y^3 + z^4$ at $(1, 2, 3)$.

$$\nabla f(x, y, z) = \frac{\partial f}{\partial x}\mathbf{i} + \frac{\partial f}{\partial y}\mathbf{j} + \frac{\partial f}{\partial z}\mathbf{k}$$

$$f(x, y, z) = x^2 + y^3 + z^4$$

$$\frac{\partial f}{\partial x} = 2x$$

$$\frac{\partial f}{\partial y} = 3y^2$$

$$\frac{\partial f}{\partial z} = 4z^3$$

$$\nabla f(x, y, z) = 2x\mathbf{i} + 3y^2\mathbf{j} + 4z^3\mathbf{k}$$

$$\nabla f(1, 2, 3) = 2\mathbf{i} + 12\mathbf{j} + 108\mathbf{k}.$$

6.7.2 Normal Vector to an Ellipse

Calculate four normal vectors for an ellipse.

The equation of an ellipse is:

$$\frac{x^2}{a^2} + \frac{y^2}{b^2} = 1.$$

Therefore,

$$f(x, y) = \frac{x^2}{a^2} + \frac{y^2}{b^2} - 1$$

$$\nabla f = \frac{2x}{a^2}\mathbf{i} + \frac{2y}{b^2}\mathbf{j}.$$

Substituting $a = 2$, $b = 1.5$ and $(x, y) = (2, 0)$, $(0, 1.5)$, $(-2, 0)$, $(0, -1.5)$:

$$\nabla f(2, 0) = \tfrac{4}{4}\mathbf{i} + \tfrac{0}{2.25}\mathbf{j}$$
$$\mathbf{n}(2, 0) = 1\mathbf{i} + 0\mathbf{j}$$
$$\nabla f(0, 1.5) = \tfrac{0}{4}\mathbf{i} + \tfrac{3}{2.25}\mathbf{j}$$
$$\mathbf{n}(0, 1.5) = 0\mathbf{i} + 1\mathbf{j}$$
$$\nabla f(-2, 0) = \tfrac{-4}{4}\mathbf{i} + \tfrac{0}{2.25}\mathbf{j}$$
$$\mathbf{n}(2, 0) = -1\mathbf{i} + 0\mathbf{j}$$
$$\nabla f(0, -1.5) = \tfrac{0}{4}\mathbf{i} - \tfrac{3}{2.25}\mathbf{j}$$
$$\mathbf{n}(0, -1.5) = 0\mathbf{i} - 1\mathbf{j}.$$

Shown in Fig. 6.21.

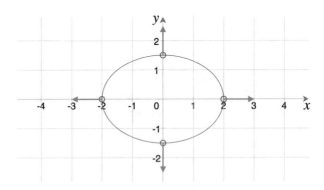

Fig. 6.21 Four normal vectors for an ellipse

6.7.3 Divergence of a Vector Field

Calculate the divergence of the vector field $\mathbf{F}(x, y, z) = x^2\mathbf{i} + y^3\mathbf{j} + z^4\mathbf{k}$, at $(1, 2, 1)$.

$$\text{div } \mathbf{F} = \nabla \cdot \mathbf{F}$$

$$\nabla \cdot \mathbf{F} = \frac{\partial}{\partial x}\left(x^2\right) + \frac{\partial}{\partial y}\left(y^3\right) + \frac{\partial}{\partial z}\left(z^4\right)$$

$$= 2x + 3y^2 + 4z^3$$

$$\nabla \cdot \mathbf{F}(1, 2, 1) = 2 + 12 + 4 = 18.$$

6.7.4 Curl of a Vector Field

Calculate the curl of the vector field $\mathbf{F}(x, y, z) = x^2 z\mathbf{i} + xy^3\mathbf{j} + yz^4\mathbf{k}$.

$$\text{curl } \mathbf{F} = \nabla \times \mathbf{F}$$

$$\nabla \times \mathbf{F} = \begin{vmatrix} \mathbf{i} & \mathbf{j} & \mathbf{k} \\ \dfrac{\partial}{\partial x} & \dfrac{\partial}{\partial y} & \dfrac{\partial}{\partial z} \\ R & S & T \end{vmatrix}$$

$$= \left(\frac{\partial T}{\partial y} - \frac{\partial S}{\partial z}\right)\mathbf{i} - \left(\frac{\partial T}{\partial x} - \frac{\partial R}{\partial z}\right)\mathbf{j} + \left(\frac{\partial S}{\partial x} - \frac{\partial R}{\partial y}\right)\mathbf{k}$$

$$R = x^2 z, \quad S = xy^3, \quad T = yz^4$$

$$\frac{\partial T}{\partial y} = z^4, \quad \frac{\partial S}{\partial z} = 0$$

$$\left(\frac{\partial T}{\partial y} - \frac{\partial S}{\partial z}\right)\mathbf{i} = z^4\mathbf{i}$$

$$\frac{\partial T}{\partial x} = 0, \quad \frac{\partial R}{\partial z} = x^2$$

$$-\left(\frac{\partial T}{\partial x} - \frac{\partial R}{\partial z}\right)\mathbf{j} = x^2\mathbf{j}$$

$$\frac{\partial S}{\partial x} = y^3, \quad \frac{\partial R}{\partial y} = 0$$

$$\left(\frac{\partial S}{\partial x} - \frac{\partial R}{\partial y}\right)\mathbf{k} = y^3\mathbf{k}$$

$$\nabla \times \mathbf{f} = z^4\mathbf{i} + x^2\mathbf{j} + y^3\mathbf{k}.$$

6.8   Summary

This chapter has introduced three differential operators: grad, div and curl, and shown how they reveal the gradient of a scalar field, normal vectors to contours and surfaces, and the divergence and curling of a vector field.

6.8.1   Summary of Formulae

The grad or gradient differential operator ∇ is defined as:

$$\nabla = \begin{bmatrix} \dfrac{\partial}{\partial x} \\ \dfrac{\partial}{\partial y} \\ \dfrac{\partial}{\partial z} \end{bmatrix} = \frac{\partial}{\partial x}\mathbf{i} + \frac{\partial}{\partial y}\mathbf{j} + \frac{\partial}{\partial z}\mathbf{k}.$$

The gradient of a scalar field $f(x, y, z)$ is:

$$\nabla f(x, y, z) = \frac{\partial f}{\partial x}\mathbf{i} + \frac{\partial f}{\partial y}\mathbf{j} + \frac{\partial f}{\partial z}\mathbf{k}.$$

If a function $f(x, y)$ or $f(x, y, z)$ equals a constant, then a normal vector $\mathbf{n}$ is

$$\mathbf{n} = \nabla f$$

and a unit normal vector is:

$$\hat{\mathbf{n}} = \frac{\nabla f}{\|\nabla f\|}.$$

The divergence of a vector field is defined as

$$\operatorname{div} \mathbf{f} = \nabla \cdot \mathbf{f}.$$

The curl of a vector field is defined as:

$$\operatorname{curl} \mathbf{f} = \nabla \times \mathbf{f}.$$

References

1. Vince J (2020) Foundation mathematics for computer science, 2nd edn. Springer, Berlin. ISBN 978-3-030-42077-2
2. Vince J (2017) Mathematics for computer graphics, 5th edn. Springer, Berlin. ISBN 978-1-4471-7334-2
3. Vince J (2019) Calculus for computer graphics, 2nd edn. Springer, Berlin. ISBN 978-3-030-11375-9

Chapter 7
Tangent and Normal Vectors

7.1 Introduction

In the previous chapter we discovered how to compute surface normal vectors using
the grad differential operator. This chapter continues this work, showing how normal
and tangent vectors are computed for a line, parabola, circle, ellipse, sine curve, cosh
curve, helix, Bézier curve, bilinear patch, quadratic Bézier patch, sphere and a torus.

7.2 Tangent Vector to a Curve

We know that the derivative of a function measures the rate of change of the function
with respect to some parameter. In terms of the function's graph, the derivative is the
slope of the graph at a point. For instance, the function $y(x) = x^3$, the first derivative
is $y'(x) = 3x^2$, as shown in Fig. 7.1. The derivative is also the slope of the *tangent
vector*, whose magnitude and direction depend upon the form of parameterisation
used for the function. For example, defining a cubic as

$$\mathbf{r}(t) = t\mathbf{i} + t^3\mathbf{j}$$

the tangent vector is

$$\frac{d\mathbf{r}}{dt} = \mathbf{r}'(t) = \mathbf{i} + 3t^2\mathbf{j}$$

whose magnitude is

$$\|\mathbf{r}'(t)\| = \sqrt{(1)^2 + (3t^2)^2} = \sqrt{1 + 9t^4}.$$

Figure 7.2 shows the cubic curve, with five tangent vectors for $t = -0.75, -0.5$,
0.0, 0.5, 0.75, which reflect the slope of the curve at the five points. However, in

© Springer-Verlag London Ltd., part of Springer Nature 2021
J. Vince, *Vector Analysis for Computer Graphics*,
https://doi.org/10.1007/978-1-4471-7505-6_7

Fig. 7.1 The graphs of
$y = x^3$, (blue) and $y' = 3x^2$,
(green) its derivative

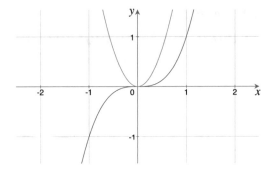

Fig. 7.2 The graph of
$y = x^3$, and five tangent
vectors

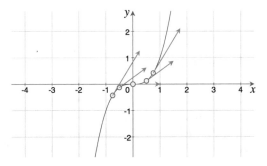

definitions for curvature, a *unit tangent vector* is important, and requires dividing
the tangent vector by its magnitude:

$$\mathbf{T}(t) = \frac{\mathbf{r}'(t)}{||\mathbf{r}'(t)||}.$$

$\mathbf{T}(t)$ is defined, only if $\mathbf{r}'(t) \neq \mathbf{0}$.

The rate of change of the unit tangent vector gives the curvature $\kappa(t)$ at any point
along the curve length s:

$$\kappa(t) = \frac{d\mathbf{T}}{ds}.$$

Figure 7.3 shows the cubic curve with five unit tangent vectors.

Generally, for a vector-valued function $\mathbf{r}(t)$, that is continuously differentiable:

$$\mathbf{r}(t) = \begin{bmatrix} x(t) \\ y(t) \end{bmatrix} \in \mathbb{R}^2, \qquad \mathbf{r}(t) = \begin{bmatrix} x(t) \\ y(t) \\ z(t) \end{bmatrix} \in \mathbb{R}^3$$

Fig. 7.3 The graph of
$y = x^3$, and five unit tangent
vectors

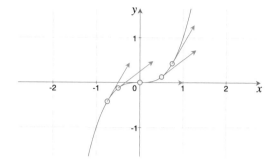

its tangent vector is

$$\frac{d\mathbf{r}}{dt} = \mathbf{r}'(t) = \begin{bmatrix} x'(t) \\ y'(t) \end{bmatrix} \neq \mathbf{0}, \qquad \frac{d\mathbf{r}}{dt} = \mathbf{r}'(t) = \begin{bmatrix} x'(t) \\ y'(t) \\ z'(t) \end{bmatrix} \neq \mathbf{0}.$$

For example, a constant pitch helix with radius ρ, is defined as

$$\mathbf{r}(t) = \begin{bmatrix} \rho \cos t \\ \rho \sin t \\ ct \end{bmatrix} = \rho \cos t \mathbf{i} + \rho \sin t \mathbf{j} + ct \mathbf{k}.$$

Therefore, its tangent vector is

$$\mathbf{r}'(t) = \begin{bmatrix} -\rho \sin t \\ \rho \cos t \\ c \end{bmatrix} = -\rho \sin t \mathbf{i} + \rho \cos t \mathbf{j} + c \mathbf{k}.$$

7.3 Normal Vector to a Curve

Ideally, a *normal vector* is orthogonal to a curve or surface, and orthogonal to its associated tangent vector. However, it would useful to confirm this mathematically. Once again, we are interested in the unit form, denoted by $\mathbf{N}(t)$.

By definition:

$$\|\mathbf{T}(t)\| = 1$$

therefore,

$$\|\mathbf{T}(t)\|^2 = 1$$

and as the dot product $\mathbf{T}(t) \cdot \mathbf{T}(t) = 1$

$$||\mathbf{T}(t)||^2 = \mathbf{T}(t) \cdot \mathbf{T}(t) = 1. \tag{7.1}$$

Differentiating (7.1), and bearing in mind that the dot product is commutative, we get

$$\frac{d}{dt}[\mathbf{T}(t) \cdot \mathbf{T}(t)] = \mathbf{T}'(t) \cdot \mathbf{T}(t) + \mathbf{T} \cdot \mathbf{T}'(t)$$

$$= 2\mathbf{T}'(t) \cdot \mathbf{T}(t) = 0.$$

For $\mathbf{T}'(t) \cdot \mathbf{T}(t) = 0$, $\mathbf{T}'(t)$ must be orthogonal to $\mathbf{T}(t)$, or $\mathbf{T}'(t) = 0$.

Thus we can define $\mathbf{N}(t)$ as

$$\mathbf{N}(t) = \frac{\mathbf{T}'(t)}{||\mathbf{T}'(t)||}.$$

Also, given a tangent vector $\mathbf{T}(t)$:

$$\mathbf{T}(t) = \begin{bmatrix} \lambda_1 \\ \lambda_2 \end{bmatrix} = \lambda_1 \mathbf{i} + \lambda_2 \mathbf{j}$$

then two vectors exist, perpendicular to $\mathbf{T}(t)$:

$$\mathbf{N}_a = \begin{bmatrix} -\lambda_2 \\ \lambda_1 \end{bmatrix} = -\lambda_2 \mathbf{i} + \lambda_1 \mathbf{j}, \quad \text{or} \quad \mathbf{N}_b = \begin{bmatrix} \lambda_2 \\ -\lambda_1 \end{bmatrix} = \lambda_2 \mathbf{i} - \lambda_1 \mathbf{j}$$

as the dot product $\mathbf{N}_a \cdot \mathbf{T}(t) = \mathbf{N}_b \cdot \mathbf{T}(t) = 0$, which means that $\mathbf{N}_a$ and $\mathbf{N}_b$ are normal vectors. Furthermore, if $\mathbf{T}(t)$ is a unit vector, so too, are $\mathbf{N}_a$ and $\mathbf{N}_b$.

But which one should we choose? Figure 7.4 shows a convention, where we see the unit normal vectors directed towards the zone containing the centre of curvature. This is called the *principal normal vector*. Another convention is to place the normal vector on one's right-hand side whilst traversing the curve.

Fig. 7.4 The graph of $y = x^3$, with unit tangent vectors (green), and unit normal vectors (red)

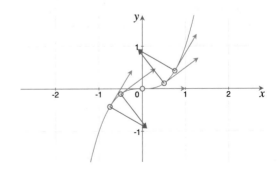

You will notice from Fig. 7.4 that there is no normal vector when $t = 0$. Let's see why.

$$\mathbf{r}'(t) = 1\mathbf{i} + 3t^2\mathbf{j}$$
$$\mathbf{r}'(0) = 1\mathbf{i} + 0\mathbf{j}$$
$$||\mathbf{r}'(t)|| = \sqrt{1 + 9t^4}$$
$$||\mathbf{r}'(0)|| = 1$$
$$\mathbf{T}(0) = 1\mathbf{i} + 0\mathbf{j}$$
$$\mathbf{T}'(0) = 0.$$

So here is a case when $\mathbf{T}'(t) = 0$.

7.3.1 Unit Tangent and Normal Vectors to a Line

Figure 7.5 shows the geometry for a parametric line, where $P_1(x_1, y_1)$ and $P(x, y)$ are two points on the line, and vector $\mathbf{s}$ provides the line's direction. Let's define $\mathbf{r}_1, \mathbf{s}, \mathbf{r}(t)$:

$$\mathbf{r}_1 = x_1\mathbf{i} + y_1\mathbf{j}$$
$$\mathbf{s} = x_s\mathbf{i} + y_s\mathbf{j}$$
$$\mathbf{r}(t) = \mathbf{r}_1 + t\mathbf{s}$$
$$= (x_1 + x_s t)\mathbf{i} + (y_1 + y_s t)\mathbf{j}.$$

Differentiating $\mathbf{r}(t)$:

$$\mathbf{r}'(t) = x_s\mathbf{i} + y_s\mathbf{j}$$

whose magnitude is

$$||\mathbf{r}'(t)|| = \sqrt{x_s^2 + y_s^2}.$$

Fig. 7.5 Geometry for a parametric line

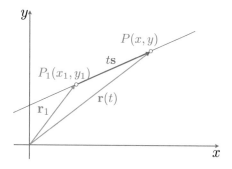

Fig. 7.6 A unit tangent and
normal vector to a line

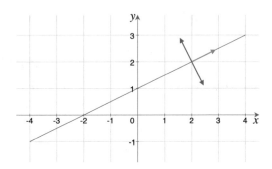

Therefore,

$$\mathbf{T} = \frac{x_s\mathbf{i} + y_s\mathbf{j}}{\sqrt{x_s^2 + y_s^2}}.$$

Figure 7.6 shows the graph of

$$\mathbf{r}(t) = 2t\mathbf{i} + (1 + t)\mathbf{j}.$$

Therefore,

$$\mathbf{T} = \frac{2\mathbf{i} + \mathbf{j}}{\sqrt{5}} \approx 0.8944\mathbf{i} + 0.4472\mathbf{j}$$

as shown in Fig. 7.6.

Differentiating **T** gives a zero vector, therefore our definition of **N** can't be used. So I'll use one of the two options described above for a perpendicular vector:

$$\mathbf{N} = -0.4472\mathbf{i} + 0.8944\mathbf{j}$$

as shown in Fig. 7.6.

Using the grad operator, we can find the unit normal vector using the line equation:

$$0 = y_s(x - x_1) - x_s(y - y_1)$$
$$f(x, y) = y_s x - x_s y - y_s x_1 + x_s y_1$$
$$\nabla f = y_s\mathbf{i} - x_s\mathbf{j}$$
$$\mathbf{N} = \frac{y_s\mathbf{i} - x_s\mathbf{j}}{\sqrt{y_s^2 + x_s^2}}.$$

Evaluating **N** for $x_s = 2$, $y_s = 1$:

$$\mathbf{N} = \frac{1\mathbf{i} - 2\mathbf{j}}{\sqrt{1 + 4}} \approx 0.4472\mathbf{i} - 0.8944\mathbf{j}$$

as shown in Fig. 7.6.

For a line in $\mathbb{R}^3$:

$$\mathbf{r}(t) = (x_1 + x_s t)\mathbf{i} + (y_1 + y_s t)\mathbf{j} + (z_1 + z_s t)\mathbf{k}$$
$$\mathbf{r}'(t) = x_s \mathbf{i} + y_s \mathbf{j} + z_s \mathbf{k}$$
$$\mathbf{T} = \frac{x_s \mathbf{i} + y_s \mathbf{j} + z_s \mathbf{k}}{\sqrt{x_s^2 + y_s^2 + z_s^2}}.$$

However, there is no unique normal vector, only a normal plane.

7.3.2 Unit Tangent and Normal Vectors to a Parabola

We normally write a parabolic equation as

$$y = ax^2 + bx + c$$

where for different values of x there is a corresponding value of y, which describes the familiar parabolic curve. However, we require this to be described as a vector-valued function. Working in two dimensions, I will align the x-component with the $\mathbf{i}$ unit vector, and the y-component with the $\mathbf{j}$ unit vector, and use a parameter t to drive the entire process. Thus it will take the general form

$$\mathbf{r}(t) = dt\mathbf{i} + \left(at^2 + bt + c\right)\mathbf{j}$$

with suitable values for a, b, c, d. Therefore, consider the parabola:

$$\mathbf{r}(t) = 2t\mathbf{i} + \left(1.5 - 1.5t^2\right)\mathbf{j}, \quad t \in [-1, 1].$$

Differentiating $\mathbf{r}(t)$:
$$\mathbf{r}'(t) = 2\mathbf{i} - 3t\mathbf{j}$$

whose magnitude is
$$||\mathbf{r}'(t)|| = \sqrt{4 + 9t^2}.$$

Therefore,
$$\mathbf{T}(t) = \frac{2\mathbf{i} - 3t\mathbf{j}}{\sqrt{4 + 9t^2}}.$$

Fig. 7.7 Three unit tangent
and normal vectors for a
parabola

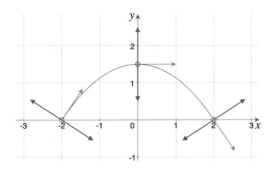

Evaluating $\mathbf{T}(t)$ for $t = -1,\ 0,\ 1$, we get:

$$\mathbf{T}(-1) = \frac{2\mathbf{i} + 3\mathbf{j}}{\sqrt{13}} \approx 0.555\mathbf{i} + 0.832\mathbf{j}$$

$$\mathbf{T}(0) = \frac{2\mathbf{i}}{\sqrt{4}} = 1\mathbf{i} + 0\mathbf{j}$$

$$\mathbf{T}(1) = \frac{2\mathbf{i} - 3\mathbf{j}}{\sqrt{13}} \approx 0.555\mathbf{i} - 0.832\mathbf{j}$$

as shown in Fig. 7.7.

Computing $\mathbf{T}'(t)$ and normalising will be rather messy, so I'll choose one of two perpendicular vectors. So given the following unit tangent vectors:

$$\mathbf{T}(-1) \approx 0.555\mathbf{i} + 0.832\mathbf{j}$$
$$\mathbf{T}(0) = 1\mathbf{i} + 0\mathbf{j}$$
$$\mathbf{T}(1) \approx 0.555\mathbf{i} - 0.832\mathbf{j}$$
$$\mathbf{N}(-1) \approx 0.832\mathbf{i} - 0.555\mathbf{j}$$
$$\mathbf{N}(0) = 0\mathbf{i} - 1\mathbf{j}$$
$$\mathbf{N}(1) \approx -0.832\mathbf{i} - 0.555\mathbf{j}$$

which point in the direction of the principal normal vector, as shown in Fig. 7.7.

Using the grad operator, we can find the unit normal vector as follows:

$$x = 2t$$
$$t = \tfrac{1}{2}x$$
$$y = 1.5 - 1.5t^2$$
$$= 1.5 - 1.5\frac{x^2}{4}$$
$$f(x, y) = y + 1.5\frac{x^2}{4} - 1.5$$

$$\nabla f = \tfrac{3}{4}x\mathbf{i} + 1\mathbf{j}$$

$$\mathbf{N}(x, y) = \frac{\tfrac{3}{4}x\mathbf{i} + 1\mathbf{j}}{\sqrt{\tfrac{9}{16}x^2 + 1}}.$$

Evaluating $\mathbf{N}(x, y)$ for three different positions:

$$\mathbf{N}(-2, 0) = \frac{\tfrac{-6}{4}\mathbf{i} + 1\mathbf{j}}{\sqrt{\tfrac{9}{16}4 + 1}} = \frac{-1.5\mathbf{i} + 1\mathbf{j}}{\sqrt{3.25}} \approx -0.832\mathbf{i} + 0.555\mathbf{j}$$

$$\mathbf{N}(0, 1.5) = \frac{0\mathbf{i} + 1\mathbf{j}}{\sqrt{1}} = 0\mathbf{i} + 1\mathbf{j}$$

$$\mathbf{N}(2, 0) = \frac{\tfrac{6}{4}\mathbf{i} + 1\mathbf{j}}{\sqrt{\tfrac{9}{16}4 + 1}} = \frac{1.5\mathbf{i} + 1\mathbf{j}}{\sqrt{3.25}} \approx 0.832\mathbf{i} + 0.555\mathbf{j}$$

as shown in Fig. 7.7.

7.3.3 Unit Tangent and Normal Vectors to a Circle

Let's find the function describing the tangent vector to a circle. We start with the following definition for a circle:

$$\mathbf{r}(t) = r \cos t\mathbf{i} + r \sin t\mathbf{j}, \quad t \in [0, 2\pi].$$

Differentiating $\mathbf{r}(t)$:

$$\mathbf{r}'(t) = -r \sin t\mathbf{i} + r \cos t\mathbf{j}$$

whose magnitude is

$$||\mathbf{r}'(t)|| = \sqrt{(-r \sin t)^2 + (r \cos t)^2} = r.$$

Note that the magnitude of the tangent vector remains constant at the circle's radius r. Therefore,

$$\mathbf{T}(t) = \frac{\mathbf{r}'(t)}{r} = -\sin t\mathbf{i} + \cos t\mathbf{j}.$$

Evaluating $\mathbf{T}(t)$ for four values of t:

$$\mathbf{T}(0°) = 0\mathbf{i} + 1\mathbf{j}$$
$$\mathbf{T}(90°) = -1\mathbf{i} + 0\mathbf{j}$$
$$\mathbf{T}(180°) = 0\mathbf{i} - 1\mathbf{j}$$

Fig. 7.8 Four unit tangent
and normal vectors for a
circle

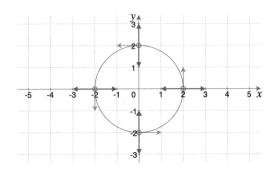

$$\mathbf{T}(270°) = 1\mathbf{i} + 0\mathbf{j}$$

as shown in Fig. 7.8.

To find $\mathbf{N}(t)$ we differentiate $\mathbf{T}(t) = -\sin t\mathbf{i} + \cos t\mathbf{j}$:

$$\mathbf{N}(t) = \mathbf{T}'(t) = -\cos t\mathbf{i} - \sin t\mathbf{j}.$$

Evaluating $\mathbf{N}(t)$ for four values of t:

$$\mathbf{N}(0°) = -1\mathbf{i} + 0\mathbf{j}$$
$$\mathbf{N}(90°) = 0\mathbf{i} - 1\mathbf{j}$$
$$\mathbf{N}(180°) = 1\mathbf{i} + 0\mathbf{j}$$
$$\mathbf{N}(270°) = 0\mathbf{i} + 1\mathbf{j}$$

as shown in Fig. 7.8.

Using the grad operator, we can find the unit normal vector as follows:

$$x^2 + y^2 = r^2$$
$$f(x, y) = x^2 + y^2 - r^2$$
$$\nabla f = 2x\mathbf{i} + 2y\mathbf{j}$$
$$\mathbf{N}(x, y) = \frac{2x\mathbf{i} + 2y\mathbf{j}}{\sqrt{4x^2 + 4y^2}}.$$

Evaluating $\mathbf{N}(x, y)$ for the same positions before, where $r = 1$:

$$\mathbf{N}(1, 0) = \frac{2\mathbf{i} + 0\mathbf{j}}{\sqrt{4}} = 1\mathbf{i} + 0\mathbf{j}$$
$$\mathbf{N}(0, 1) = \frac{0\mathbf{i} + 3\mathbf{j}}{\sqrt{4}} = 0\mathbf{i} + 1\mathbf{j}$$
$$\mathbf{N}(-1, 0) = \frac{-2\mathbf{i} + 0\mathbf{j}}{\sqrt{4}} = -1\mathbf{i} + 0\mathbf{j}$$

$$\mathbf{N}(0, -1) = \frac{0\mathbf{i} - 2\mathbf{j}}{\sqrt{4}} = 0\mathbf{i} - 1\mathbf{j}$$

as shown in Fig. 7.8.

7.3.4 Unit Tangent and Normal Vectors to an Ellipse

Having found the unit tangent and normal vectors for a circle, an ellipse should be similar. Let's define an ellipse as

$$\mathbf{r}(t) = a \cos t\mathbf{i} + b \sin t\mathbf{j}, \quad t \in [0, 2\pi].$$

Differentiating $\mathbf{r}(t)$:

$$\mathbf{r}'(t) = -a \sin t\mathbf{i} + b \cos t\mathbf{j}$$

whose magnitude is

$$
\begin{aligned}
||\mathbf{r}'(t)|| &= \sqrt{\left(-a \sin t\right)^2 + \left(b \cos t\right)^2} \\
&= \sqrt{a^2 \sin^2 t + b^2 \cos^2 t} \\
&= \sqrt{a^2\left(1 - \cos^2 t\right) + b^2 \cos^2 t} \\
&= \sqrt{a^2 - \left(a^2 - b^2\right) \cos^2 t} \\
&= a\sqrt{1 - \epsilon^2 \cos^2 t}
\end{aligned}
$$

where $\epsilon = \sqrt{1 - b^2/a^2}$ is the eccentricity of the ellipse.
Therefore,

$$\mathbf{T}(t) = \frac{-a \sin t\mathbf{i} + b \cos t\mathbf{j}}{a\sqrt{1 - \epsilon^2 \cos^2 t}}.$$

As an example, let's define an ellipse with $a = 2$ and $b = 1.5$, which makes the eccentricity:

$$\epsilon = \sqrt{1 - 1.5^2/2^2} = \sqrt{0.4375}.$$

Evaluating $\mathbf{T}(t)$ for four values of t:

$$\mathbf{T}(0°) = \frac{0\mathbf{i} + 1.5\mathbf{j}}{2\sqrt{1 - 0.4375}} = 0\mathbf{i} + 1\mathbf{j}$$

$$\mathbf{T}(90°) = -\frac{-2\mathbf{i} + 0\mathbf{j}}{2\sqrt{1}} = -1\mathbf{i} + 0\mathbf{j}$$

Fig. 7.9 Four unit tangent
and normal vectors for an
ellipse

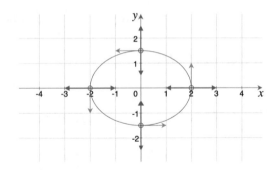

$$T(180°) = \frac{0i - 1.5j}{2\sqrt{1 - 0.4375}} = 0i - 1j$$

$$T(270°) = \frac{2i + 0j}{2\sqrt{1}} = 1i + 0j$$

as shown in Fig. 7.9.

Once again, there is no need to differentiate $T(t)$ to find $N(t)$. We simply use the
perpendicular strategy explained above. Therefore,

$$N(0°) = -1i + 0j$$
$$N(90°) = 0i - 1j$$
$$N(180°) = 1i + 0j$$
$$N(270°) = 0i + 1j$$

as shown in Fig. 7.9.

Using the grad operator, we can find the unit normal vector as follows:

$$\frac{x^2}{a^2} + \frac{y^2}{b^2} = 1$$

$$f(x, y) = \frac{x^2}{a^2} + \frac{y^2}{b^2} - 1$$

$$\nabla f = \frac{2x}{a^2}i + \frac{2y}{b^2}j.$$

Substituting $a = 2$, $b = 1.5$ and $(x, y) = (2, 0)$, $(0, 1.5)$, $(-2, 0)$, $(0, -1.5)$:

$$\nabla f(2, 0) = \tfrac{4}{4}i + \tfrac{0}{2.25}j$$
$$N(2, 0) = 1i + 0j$$
$$\nabla f(0, 1.5) = \tfrac{0}{4}i + \tfrac{3}{2.25}j$$
$$N(0, 1.5) = 0i + 1j$$
$$\nabla f(-2, 0) = \tfrac{-4}{4}i + \tfrac{0}{2.25}j$$

$$\mathbf{N}(2, 0) = -1\mathbf{i} + 0\mathbf{j}$$
$$\nabla f(0, -1.5) = \tfrac{0}{4}\mathbf{i} - \tfrac{3}{2.25}\mathbf{j}$$
$$\mathbf{N}(0, -1.5) = 0\mathbf{i} - 1\mathbf{j}$$

as shown in Fig. 7.9.

7.3.5 Unit Tangent and Normal Vectors to a Sine Curve

Let's calculate the tangent and normal vectors to a sine waveform, which may be of interest in the rendering of sinusoidal waves. We define one period of a sine curve as

$$\mathbf{r}(t) = t\mathbf{i} + 2\sin t\mathbf{j}, \quad t \in [0, 2\pi].$$

Differentiating $\mathbf{r}(t)$:

$$\mathbf{r}'(t) = 1\mathbf{i} + 2\cos t\mathbf{j}$$

whose magnitude is

$$\|\mathbf{r}'(t)\| = \sqrt{1 + 4\cos^2 t}.$$

Therefore,

$$\mathbf{T}(t) = \frac{\mathbf{i} + 2\cos t\mathbf{j}}{\sqrt{1 + 4\cos^2 t}}.$$

Evaluating $\mathbf{T}(t)$ for four values of t:

$$\mathbf{T}(0°) = \frac{1\mathbf{i} + 2\mathbf{j}}{\sqrt{5}} \approx 0.4472\mathbf{i} + 0.8944\mathbf{j}$$
$$\mathbf{T}(90°) = \frac{1\mathbf{i} + 0\mathbf{j}}{\sqrt{1}} = 1\mathbf{i} + 0\mathbf{j}$$
$$\mathbf{T}(180°) = \frac{1\mathbf{i} - 2\mathbf{j}}{\sqrt{5}} \approx 0.4472\mathbf{i} - 0.8944\mathbf{j}$$
$$\mathbf{T}(270°) = \frac{1\mathbf{i} + 0\mathbf{j}}{\sqrt{1}} = 1\mathbf{i} + 0\mathbf{j}$$

as shown in Fig. 7.10.

To differentiate $\mathbf{T}(t)$ and normalise it, looks as though it requires considerable work, so we'll take the easy root as before. Therefore,

$$\mathbf{N}(0°) \approx 0.8944\mathbf{i} - 0.4472\mathbf{j}$$
$$\mathbf{N}(90°) = 0\mathbf{i} - 1\mathbf{j}$$
$$\mathbf{N}(180°) \approx -0.8944\mathbf{i} - 0.4472\mathbf{j}$$

Fig. 7.10 Four unit tangent
and normal vectors for a sine
curve

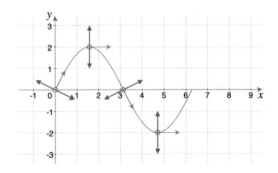

$$N(270°) = 0i + 1j$$

as shown in Fig. 7.10.

Using the grad operator, we can find the unit normal vector as follows:

$$y = 2\sin x$$
$$f(x, y) = y - 2\sin x = 0$$
$$\nabla f = -2\cos x\, i + 1j$$
$$N(x) = \frac{-2\cos x\, i + 1j}{\sqrt{1 + 4\cos^2 x}}.$$

Evaluating $N(x)$ for four values of x:

$$N(0) = \frac{-2i + 1j}{\sqrt{5}} \approx -0.8944i + 0.4472j$$
$$N(\pi/2) = 0i + 1j$$
$$N(\pi) \approx 0.8944i + 0.4472j$$
$$N(3\pi/2) = 0i - 1j$$

as shown in Fig. 7.10.

7.3.6 Unit Tangent and Normal Vectors to a **cosh** Curve

Now let's calculate the tangent and normal vectors to a cosh curve, also called a
catenary. We define part of a cosh curve as

$$r(t) = ti + 3\cosh\left(\tfrac{t}{3}\right) j, \quad t \in [-3, 3].$$

Fig. 7.11 Three unit tangent and normal vectors for $3\cosh(x/3)$

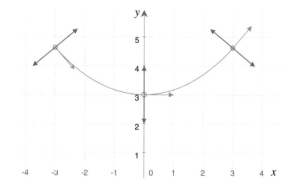

Differentiating $\mathbf{r}(t)$:

$$\mathbf{r}'(t) = 1\mathbf{i} + \sinh\left(\tfrac{t}{3}\right)\mathbf{j}$$

whose magnitude is

$$\|\mathbf{r}'(t)\| = \sqrt{1 + \sinh^2\left(\tfrac{t}{3}\right)} = \cosh\left(\tfrac{t}{3}\right).$$

Therefore,

$$\mathbf{T}(t) = \frac{1\mathbf{i} + \sinh\left(\tfrac{t}{3}\right)\mathbf{j}}{\cosh\left(\tfrac{t}{3}\right)}.$$

Let's find $\mathbf{T}(t)$ for three values of t:

$$\mathbf{T}(-3) = \frac{1\mathbf{i} - 1.1752\mathbf{j}}{1.5431} \approx 0.6481\mathbf{i} - 0.7616\mathbf{j}$$

$$\mathbf{T}(0) = 1\mathbf{i} + 0\mathbf{j}$$

$$\mathbf{T}(3) = \frac{1\mathbf{i} + 1.1752\mathbf{j}}{1.5431} \approx 0.6481\mathbf{i} + 0.7616\mathbf{j}$$

as shown in Fig. 7.11.

This time, let's differentiate $\mathbf{T}(t)$ and normalise it:

$$\mathbf{T}(t) = \frac{1\mathbf{i} + \sinh\left(\tfrac{t}{3}\right)\mathbf{j}}{\cosh\left(\tfrac{t}{3}\right)} = \mathrm{sech}\left(\tfrac{t}{3}\right)\mathbf{i} + \tanh\left(\tfrac{t}{3}\right)\mathbf{j}$$

$$\mathbf{T}'(t) = -\tfrac{1}{3}\left(\mathrm{sech}\left(\tfrac{t}{3}\right)\tanh\left(\tfrac{t}{3}\right)\mathbf{i} + \mathrm{sech}^2\left(\tfrac{t}{3}\right)\mathbf{j}\right)$$

$$= -\tfrac{1}{3}\left(\frac{\sinh\left(\tfrac{t}{3}\right)}{\cosh^2\left(\tfrac{t}{3}\right)}\mathbf{i} + \frac{1}{\cosh^2\left(\tfrac{t}{3}\right)}\mathbf{j}\right)$$

$$\|\mathbf{T}'(t)\| = \tfrac{1}{3}\sqrt{\frac{\sinh^2\left(\tfrac{t}{3}\right)}{\cosh^4\left(\tfrac{t}{3}\right)} + \frac{1}{\cosh^4\left(\tfrac{t}{3}\right)}} = \tfrac{1}{3}\mathrm{sech}\left(\tfrac{t}{3}\right)$$

$$\mathbf{N}(t) = -\cosh\left(\tfrac{t}{3}\right)\left(\frac{\sinh\left(\tfrac{t}{3}\right)}{\cosh^2\left(\tfrac{t}{3}\right)}\mathbf{i} + \frac{1}{\cosh^2\left(\tfrac{t}{3}\right)}\mathbf{j}\right)$$

$$= -\tanh\left(\tfrac{t}{3}\right)\mathbf{i} + \operatorname{sech}\left(\tfrac{t}{3}\right)\mathbf{j}.$$

Note that this is one of the options if we had taken the easy route!
Therefore,

$$\mathbf{N}(-3) \approx 0.7616\mathbf{i} + 0.6481\mathbf{j}$$
$$\mathbf{N}(0) = 0\mathbf{i} + 1\mathbf{j}$$
$$\mathbf{N}(3) \approx -0.7616\mathbf{i} + 0.6481\mathbf{j}$$

as shown in Fig. 7.11.
Using the grad operator, we can find the unit normal vector as follows:

$$y = 3\cosh\left(\tfrac{t}{3}\right)$$
$$f(x, y) = y - 3\cosh\left(\tfrac{t}{3}\right)$$
$$\nabla f = -\sinh\left(\tfrac{t}{3}\right)\mathbf{i} + 1\mathbf{j}$$
$$\mathbf{N}(x) = \frac{-\sinh\left(\tfrac{t}{3}\right)\mathbf{i} + 1\mathbf{j}}{\sqrt{1 + \sinh^2\left(\tfrac{t}{3}\right)}}$$
$$= \frac{-\sinh\left(\tfrac{t}{3}\right)\mathbf{i} + 1\mathbf{j}}{\cosh\left(\tfrac{t}{3}\right)}$$
$$= -\tanh\left(\tfrac{t}{3}\right)\mathbf{i} + \operatorname{sech}\left(\tfrac{t}{3}\right)\mathbf{j}$$

which gives the same result as before.

7.3.7   Unit Tangent and Normal Vectors to a Helix

A helix is a 3D curve and used in nature to store the genetic code of all living organisms. It can have a variable radius, and also a variable pitch. However, a fixed radius and constant-pitch helix is a popular curve used for illustrating tangent and normal vectors. Let's define a helix as

$$\mathbf{r}(t) = 2\cos t\mathbf{i} + 2\sin t\mathbf{j} + t\mathbf{k}, \quad t \in [0, 4\pi].$$

Differentiating $\mathbf{r}(t)$:

$$\mathbf{r}'(t) = -2\sin t\mathbf{i} + 2\cos t\mathbf{j} + 1\mathbf{k}$$

Fig. 7.12 Three unit tangent and normal vectors for a helix

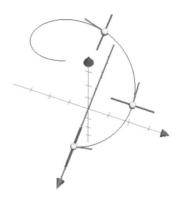

whose magnitude is

$$\|\mathbf{r}(t)\| = \sqrt{4\sin^2 t + 4\cos^2 t + 1} = \sqrt{5}.$$

Therefore,

$$\mathbf{T}(t) = \tfrac{1}{\sqrt{5}}\left(-2\sin t\mathbf{i} + 2\cos t\mathbf{j} + \mathbf{k}\right).$$

Evaluating $\mathbf{T}(t)$ for different values of t:

$$\mathbf{T}(0) = \tfrac{1}{\sqrt{5}}\left(2\mathbf{j} + 1\mathbf{k}\right) \approx 0.8944\mathbf{j} + 0.4472\mathbf{k}$$
$$\mathbf{T}(\pi/2) = \tfrac{1}{\sqrt{5}}\left(-2\mathbf{i} + 1\mathbf{k}\right) \approx -0.8944\mathbf{i} + 0.4472\mathbf{k}$$
$$\mathbf{T}(\pi) = \tfrac{1}{\sqrt{5}}\left(-2\mathbf{j} + 1\mathbf{k}\right) \approx -0.8944\mathbf{j} + 0.4472\mathbf{k}$$

as shown in Fig. 7.12.

Differentiating $\mathbf{T}(t)$ and normalising:

$$\mathbf{T}'(t) = \tfrac{1}{\sqrt{5}}\left(-2\cos t\mathbf{i} - 2\sin t\mathbf{j}\right)$$
$$\|\mathbf{T}'(t)\| = \tfrac{1}{\sqrt{5}}\sqrt{4\cos^2 t + 4\sin^2 t} = \tfrac{2}{\sqrt{5}}$$
$$\mathbf{N}(t) = \frac{\tfrac{1}{\sqrt{5}}\left(-2\cos t\mathbf{i} - 2\sin t\mathbf{j}\right)}{\tfrac{2}{\sqrt{5}}} = -\cos t\mathbf{i} - \sin t\mathbf{j}.$$

Evaluating $\mathbf{N}(t)$ for different values of t:

$$\mathbf{N}(0) = -1\mathbf{i} + 0\mathbf{j}$$
$$\mathbf{N}(\pi/2) = 0\mathbf{i} - 1\mathbf{j}$$
$$\mathbf{N}(t) = 1\mathbf{i} + 0\mathbf{j}$$

as shown in Fig. 7.12.

7.3.8 Unit Tangent and Normal Vectors to a Quadratic Bézier Curve

Quadratic curves are normally expressed using a basis function $\mathbf{B}(t)$, which generates values using the parameter t. I will derive the derivative of a general basis function. A 2-D quadratic curve is expressed using a column vector as

$$\mathbf{r}(t) = \begin{bmatrix} x(t) \\ y(t) \end{bmatrix}, \quad t \in [0, \ 1]$$

$$x(t) = \mathbf{B}_{2,0}(t)x_0 + \mathbf{B}_{2,1}(t)x_1 + \mathbf{B}_{2,2}(t)x_2$$

$$y(t) = \mathbf{B}_{2,0}(t)y_0 + \mathbf{B}_{2,1}(t)y_1 + \mathbf{B}_{2,2}(t)y_2$$

$$\mathbf{B}_{2,0} = (1 - t)^2$$

$$\mathbf{B}_{2,1} = 2t(1 - t)$$

$$\mathbf{B}_{2,2} = t^2.$$

Algebraically:

$$\mathbf{r}(t) = \mathbf{B}_{2,0}(t)\mathbf{P}_0 + \mathbf{B}_{2,1}(t)\mathbf{P}_1 + \mathbf{B}_{2,2}(t)\mathbf{P}_2, \quad t \in [0, 1]$$

where $\mathbf{P}_0, \ \mathbf{P}_1, \ \mathbf{P}_2$ are position vectors for the control point $P_0, \ P_1, \ P_2$.

But a Cartesian vector is rarely used. So, for the time being, I will use algebraic notation.

Let's start with the following 2-D quadratic Bézier curve:

$$\mathbf{r}(t) = \mathbf{P}_0(1 - t)^2 + 2\mathbf{P}_1 t(1 - t) + \mathbf{P}_2 t^2, \quad t \in [0, 1].$$

Differentiating $\mathbf{r}(t)$:

$$\begin{aligned} \mathbf{r}'(t) &= -2\mathbf{P}_0(1 - t) + 2\mathbf{P}_1(1 - 2t) + 2\mathbf{P}_2 t \\ &= -2\mathbf{P}_0 + 2\mathbf{P}_0 t + 2\mathbf{P}_1 - 4\mathbf{P}_1 t + 2\mathbf{P}_2 t \\ &= 2(\mathbf{P}_1 - \mathbf{P}_0)(1 - t) + 2(\mathbf{P}_2 - \mathbf{P}_1)t \\ &= 2\big((\mathbf{P}_1 - \mathbf{P}_0)(1 - t) + (\mathbf{P}_2 - \mathbf{P}_1)t\big) \\ x'(t) &= 2\big((x_1 - x_0)(1 - t) + (x_2 - x_1)t\big) \\ y'(t) &= 2\big((y_1 - y_0)(1 - t) + (y_2 - y_1)t\big) \end{aligned}$$

whose magnitude is

$$\|\mathbf{r}'(t)\| = \sqrt{\big(x'(t)\big)^2 + \big(y'(t)\big)^2}.$$

Fig. 7.13 Three unit tangent and normal vectors for a quadratic Bézier curve

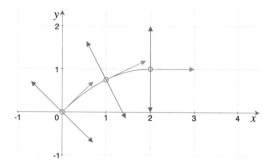

Therefore,

$$\mathbf{T}(t) = \frac{2[(\mathbf{P}_1 - \mathbf{P}_0)(1 - t) + (\mathbf{P}_2 - \mathbf{P}_1)t]}{\sqrt{(x'(t))^2 + (y'(t))^2}}.$$

Now let's substitute specific values for $\mathbf{P}_0$, $\mathbf{P}_1$, $\mathbf{P}_2$, $\mathbf{P}_0 = (0, 0)$, $\mathbf{P}_1 = (1, 1)$, $\mathbf{P}_2 = (2, 1)$. Therefore,

$$x'(t) = 2\big((1 - 0)(1 - t) + (2 - 1)t\big) = 2$$
$$y'(t) = 2\big((1 - 0)(1 - t) + (1 - 1)t\big) = 2(1 - t)$$
$$\mathbf{T}(t) = \frac{2\mathbf{i} + 2(1 - t)\mathbf{j}}{\sqrt{4 + 4(1 - t)^2}} = \frac{1\mathbf{i} + (1 - t)\mathbf{j}}{\sqrt{1 + (1 - t)^2}}.$$

Evaluating $\mathbf{T}(t)$ for different values of t:

$$\mathbf{T}(0) = \frac{1\mathbf{i} + 1\mathbf{j}}{\sqrt{2}} \approx 0.7071\mathbf{i} + 0.7071\mathbf{j}$$
$$\mathbf{T}(0.5) = \frac{1\mathbf{i} + 0.5\mathbf{j}}{\sqrt{1.25}} \approx 0.8944\mathbf{i} + 0.4472\mathbf{j}$$
$$\mathbf{T}(1) = 1\mathbf{i} + 0\mathbf{j}$$

as shown in Fig. 7.13.

Differentiating $\mathbf{T}(t)$ and normalising looks like a lot of work, so we'll take the easy route. Therefore,

$$\mathbf{N}(0) \approx 0.7071\mathbf{i} - 0.7071\mathbf{j}$$
$$\mathbf{N}(0.5) \approx 0.4472\mathbf{i} - 0.8944\mathbf{j}$$
$$\mathbf{N}(1) = 0\mathbf{i} - 1\mathbf{j}$$

as shown in Fig. 7.13.

7.4   Unit Tangent and Normal Vectors to a Surface

In the following examples I show how to calculate the tangent and normal vectors to a bilinear patch, a Bézier patch, a sphere and a torus. They each require a slightly different approach, which is explained for each surface.

7.4.1   Unit Normal Vectors to a Bilinear Patch

Bilinear patches are constructed from a pair of lines using linear interpolation. For example, given two lines defined by their position vectors:

$$L_1 = (\mathbf{P}_0, \mathbf{P}_1), \qquad L_2 = (\mathbf{P}_2, \mathbf{P}_3)$$

we can linearly interpolate along the lines using

$$\mathbf{A}(u) = (1 - u)\mathbf{P}_0 + u\mathbf{P}_1, \quad u \in [0, 1]$$
$$\mathbf{B}(u) = (1 - u)\mathbf{P}_2 + u\mathbf{P}_3, \quad u \in [0, 1]$$

and then linearly interpolate between $\mathbf{A}(u)$ and $\mathbf{B}(u)$:

$$
\begin{aligned}
\mathbf{r}(u, v) &= (1 - v)\mathbf{A}(u) + v\mathbf{B}(u), \quad v \in [0, 1] \\
&= (1 - v)\big((1 - u)\mathbf{P}_0 + u\mathbf{P}_1\big) + v\big((1 - u)\mathbf{P}_2 + u\mathbf{P}_3\big) \\
&= (1 - v)(1 - u)\mathbf{P}_0 + u(1 - v)\mathbf{P}_1 + v(1 - u)\mathbf{P}_2 + uv\mathbf{P}_3.
\end{aligned}
$$

We now compute the partial derivatives for u and v:

$$
\begin{aligned}
\frac{\partial \mathbf{r}}{\partial u} &= -(1 - v)\mathbf{P}_0 + (1 - v)\mathbf{P}_1 - v\mathbf{P}_2 + v\mathbf{P}_3 \\
&= (1 - v)(\mathbf{P}_1 - \mathbf{P}_0) + v(\mathbf{P}_3 - \mathbf{P}_2) \\
\frac{\partial \mathbf{r}}{\partial v} &= -(1 - u)\mathbf{P}_0 - u\mathbf{P}_1 + (1 - u)\mathbf{P}_2 + u\mathbf{P}_3 \\
&= (1 - u)(\mathbf{P}_2 - \mathbf{P}_0) + u(\mathbf{P}_3 - \mathbf{P}_1).
\end{aligned}
$$

$\frac{\partial \mathbf{r}}{\partial u}$ and $\frac{\partial \mathbf{r}}{\partial v}$ encode a pair of orthogonal tangent vectors, whose cross-product is a vector normal.

Let's demonstrate this with an example. Given:

$$
\begin{aligned}
\mathbf{P}_0 &= 0\mathbf{i} + 0\mathbf{j} + 0\mathbf{k} \\
\mathbf{P}_1 &= 0\mathbf{i} + 2\mathbf{j} + 1\mathbf{k} \\
\mathbf{P}_2 &= 2\mathbf{i} + 0\mathbf{j} + 0\mathbf{k}
\end{aligned}
$$

$$\mathbf{P}_3 = 2\mathbf{i} + 2\mathbf{j} - 1\mathbf{k}$$

then,

$$\frac{\partial \mathbf{r}}{\partial u} = 0\mathbf{i} + \big(2(1-v) + 2v\big)\mathbf{j} + \big((1-v) - v\big)\mathbf{k}$$

$$= 0\mathbf{i} + 2\mathbf{j} + (1 - 2v)\mathbf{k}$$

$$\frac{\partial \mathbf{r}}{\partial v} = \big(2(1-u) + 2u\big)\mathbf{i} + 0\mathbf{j} - 2u\mathbf{k}$$

$$= 2\mathbf{i} + 0\mathbf{j} - 2u\mathbf{k}.$$

We can now calculate their cross product:

$$\mathbf{T}'(u,v) = \begin{vmatrix} \mathbf{i} & \mathbf{j} & \mathbf{k} \\ 2 & 0 & -2u \\ 0 & 2 & 1-2v \end{vmatrix} = 4u\mathbf{i} + (4v - 2)\mathbf{j} + 4\mathbf{k}$$

which is a vector orthogonal to the tangent vectors, depending on the value of u and v. Let's calculate the unit normal vector by dividing the normal vector by its magnitude, for different values of u and v.

$$\mathbf{N}(0,0) = \frac{0\mathbf{i} - 2\mathbf{j} + 4\mathbf{k}}{\sqrt{20}} \approx 0\mathbf{i} - 0.4472\mathbf{j} + 0.8944\mathbf{k}$$

$$\mathbf{N}(1,0) = \frac{4\mathbf{i} - 2\mathbf{j} + 4\mathbf{k}}{\sqrt{34}} \approx 0.686\mathbf{i} - 0.343\mathbf{j} + 0.686\mathbf{k}$$

$$\mathbf{N}(0,1) = \frac{0\mathbf{i} + 2\mathbf{j} + 4\mathbf{k}}{\sqrt{20}} \approx 0\mathbf{i} + 0.4472\mathbf{j} + 0.8944\mathbf{k}$$

$$\mathbf{N}(1,1) = \frac{4\mathbf{i} + 2\mathbf{j} + 4\mathbf{k}}{\sqrt{34}} \approx 0.686\mathbf{i} + 0.343\mathbf{j} + 0.686\mathbf{k}$$

$$\mathbf{N}(0.5, 0.5) = \frac{2\mathbf{i} + 0\mathbf{j} + 4\mathbf{k}}{\sqrt{2}} \approx 0.4472\mathbf{i} + 0\mathbf{j} + 0.8944\mathbf{k}$$

as shown in Fig. 7.14

7.4.2   Unit Normal Vectors to a Quadratic Bézier Patch

Bézier proposed a matrix of nine control points to determine the geometry of a quadratic patch, as shown in Fig. 7.15. Any point on the patch is defined by

Fig. 7.14 Five unit normal
vectors for a bilinear surface

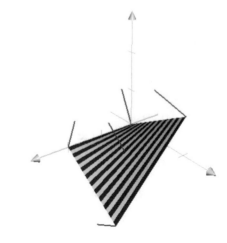

Fig. 7.15 A quadratic
Bézier surface patch

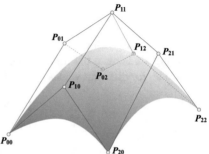

$$\mathbf{P}_{uv} = [u^2 \ \ u \ \ 1] \begin{bmatrix} 1 & -2 & 1 \\ -2 & 2 & 0 \\ 1 & 0 & 0 \end{bmatrix} \begin{bmatrix} P_{00} & P_{01} & P_{02} \\ P_{10} & P_{11} & P_{12} \\ P_{20} & P_{21} & P_{22} \end{bmatrix} \begin{bmatrix} 1 & -2 & 1 \\ -2 & 2 & 0 \\ 1 & 0 & 0 \end{bmatrix} \begin{bmatrix} v^2 \\ v \\ 1 \end{bmatrix}.$$

The individual x-, y- and z-coordinates are obtained by substituting the x, y, z
values for the central **P** matrix.

Let's illustrate the process with an example. Given the following points:

$$P_{00} = (0, \ 0, \ 0), \quad P_{01} = (1, \ 1, \ 0), \quad P_{02} = (2, \ 0, \ 0)$$
$$P_{10} = (0, \ 1, \ 1), \quad P_{11} = (1, \ 2, \ 1), \quad P_{12} = (2, \ 1, \ 1)$$
$$P_{20} = (0, \ 0, \ 2), \quad P_{21} = (1, \ 1, \ 2), \quad P_{22} = (2, \ 0, \ 2)$$

we can write:

$$x_{uv} = [u^2 \ \ u \ \ 1] \begin{bmatrix} 1 & -2 & 1 \\ -2 & 2 & 0 \\ 1 & 0 & 0 \end{bmatrix} \begin{bmatrix} 0 & 1 & 2 \\ 0 & 1 & 2 \\ 0 & 1 & 2 \end{bmatrix} \begin{bmatrix} 1 & -2 & 1 \\ -2 & 2 & 0 \\ 1 & 0 & 0 \end{bmatrix} \begin{bmatrix} v^2 \\ v \\ 1 \end{bmatrix}$$

$$x_{uv} = [u^2 \quad u \quad 1] \begin{bmatrix} 0 & 0 & 0 \\ 0 & 0 & 0 \\ 0 & 2 & 0 \end{bmatrix} \begin{bmatrix} v^2 \\ v \\ 1 \end{bmatrix}$$

$$x_{uv} = 2v,$$

$$y_{uv} = [u^2 \quad u \quad 1] \begin{bmatrix} 1 & -2 & 1 \\ -2 & 2 & 0 \\ 1 & 0 & 0 \end{bmatrix} \begin{bmatrix} 0 & 1 & 0 \\ 1 & 2 & 1 \\ 0 & 1 & 0 \end{bmatrix} \begin{bmatrix} 1 & -2 & 1 \\ -2 & 2 & 0 \\ 1 & 0 & 0 \end{bmatrix} \begin{bmatrix} v^2 \\ v \\ 1 \end{bmatrix}$$

$$y_{uv} = [u^2 \quad u \quad 1] \begin{bmatrix} 0 & 0 & -2 \\ 0 & 0 & 2 \\ -2 & 2 & 0 \end{bmatrix} \begin{bmatrix} v^2 \\ v \\ 1 \end{bmatrix}$$

$$y_{uv} = 2(u + v - u^2 - v^2),$$

$$z_{uv} = [u^2 \quad u \quad 1] \begin{bmatrix} 1 & -2 & 1 \\ -2 & 2 & 0 \\ 1 & 0 & 0 \end{bmatrix} \begin{bmatrix} 0 & 0 & 0 \\ 1 & 1 & 1 \\ 2 & 2 & 2 \end{bmatrix} \begin{bmatrix} 1 & -2 & 1 \\ -2 & 2 & 0 \\ 1 & 0 & 0 \end{bmatrix} \begin{bmatrix} v^2 \\ v \\ 1 \end{bmatrix}$$

$$z_{uv} = [u^2 \quad u \quad 1] \begin{bmatrix} 0 & 0 & 0 \\ 0 & 0 & 2 \\ 0 & 0 & 0 \end{bmatrix} \begin{bmatrix} v^2 \\ v \\ 1 \end{bmatrix}$$

$$z_{uv} = 2u.$$

Therefore, any point on the surface patch has coordinates

$$\mathbf{p}_{uv} = 2v\mathbf{i} + 2\left(u + v - u^2 - v^2\right)\mathbf{j} + 2u\mathbf{k}.$$

To calculate a unit vector normal to the surface we first calculate two tangent vectors using $\frac{\partial \mathbf{p}}{\partial u}$ and $\frac{\partial \mathbf{p}}{\partial v}$, take their cross product, and normalise the resulting vector.

$$\frac{\partial \mathbf{p}}{\partial u} = 0\mathbf{i} + 2(1 - 2u)\mathbf{j} + 2\mathbf{k}$$

$$\frac{\partial \mathbf{p}}{\partial v} = 2\mathbf{i} + 2(1 - 2v)\mathbf{j} + 0\mathbf{k}.$$

We can now compute their cross product:

$$\begin{vmatrix} \mathbf{i} & \mathbf{j} & \mathbf{k} \\ 0 & 2 - 4u & 2 \\ 2 & 2 - 4v & 0 \end{vmatrix} = (8v - 4)\mathbf{i} + 4\mathbf{j} + (8u - 4)\mathbf{k}$$

which is a vector orthogonal to the tangent vectors, depending on the value of u and v. Let's calculate the unit normal vector by dividing the normal vector by its magnitude, for different values of u and v.

$$\mathbf{N}(0, 0) = \frac{-4\mathbf{i} + 4\mathbf{j} - 4\mathbf{k}}{\sqrt{48}} \approx -0.5774\mathbf{i} + 0.5774\mathbf{j} - 0.5774\mathbf{k}$$

Fig. 7.16 Five unit normal
vectors for a quadratic
Bézier patch

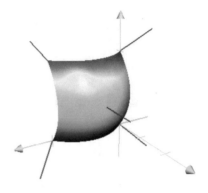

$$\mathbf{N}(1, 0) = \frac{-4\mathbf{i} + 4\mathbf{j} + 4\mathbf{k}}{\sqrt{48}} \approx -0.5774\mathbf{i} + 0.5774\mathbf{j} + 0.5774\mathbf{k}$$

$$\mathbf{N}(0, 1) = \frac{4\mathbf{i} + 4\mathbf{j} - 4\mathbf{k}}{\sqrt{48}} \approx 0.5774\mathbf{i} + 0.5774\mathbf{j} - 0.5774\mathbf{k}$$

$$\mathbf{N}(1, 1) = \frac{4\mathbf{i} + 4\mathbf{j} + 4\mathbf{k}}{\sqrt{48}} \approx 0.5774\mathbf{i} + 0.5774\mathbf{j} + 0.5774\mathbf{k}$$

$$\mathbf{N}(0.5, 0.5) = \frac{0\mathbf{i} + 4\mathbf{j} + 0\mathbf{k}}{\sqrt{16}} \approx 0\mathbf{i} + 1\mathbf{j} + 0\mathbf{k}$$

as shown in Fig. 7.16.

7.4.3 Unit Tangent and Normal Vectors to a Sphere

It should not be too difficult to find the tangent and normal vectors for a sphere. So
let's start with the equation for a sphere with radius r in Cartesian coordinates as

$$x^2 + y^2 + z^2 = r^2.$$

Therefore, we can declare a function:

$$f(x, y, z) = x^2 + y^2 + z^2 - r^2.$$

This is best solved using the notation of a gradient vector ∇.
    Therefore,

$$\frac{\partial f}{\partial x} = 2x$$

$$\frac{\partial f}{\partial y} = 2y$$

$$\frac{\partial f}{\partial z} = 2z$$

$$\nabla f = 2x\mathbf{i} + 2y\mathbf{j} + 2z\mathbf{k}$$

which is a vector normal to the sphere.

Let's compute the unit normal vector for different points on a sphere using

$$\mathbf{N} = \frac{\nabla f}{\|\nabla f\|}.$$

But in order to identify points on the sphere's surface, it is easier to use spherical coordinates, where

$$x = r \sin \phi \cos \theta$$
$$y = r \sin \phi \sin \theta$$
$$z = r \cos \phi.$$

Therefore,

$$\nabla f (r, \phi, \theta) = 2r \sin \phi \cos \theta \mathbf{i} + 2r \sin \phi \sin \theta \mathbf{j} + 2r \cos \phi \mathbf{k}$$

$$\mathbf{N} \left(5, \tfrac{\pi}{2}, 0 \right) = \frac{10\mathbf{i} + 0\mathbf{j} + 0\mathbf{k}}{\sqrt{100}} = 1\mathbf{i} + 0\mathbf{j} + 0\mathbf{k}$$

$$\mathbf{N} \left(5, \tfrac{\pi}{2}, \tfrac{\pi}{2} \right) = \frac{0\mathbf{i} + 10\mathbf{j} + 0\mathbf{k}}{\sqrt{100}} = 0\mathbf{i} + 1\mathbf{j} + 0\mathbf{k}$$

$$\mathbf{N} (5, 0, 0) = \frac{0\mathbf{i} + 0\mathbf{j} + 10\mathbf{k}}{\sqrt{100}} = 0\mathbf{i} + 0\mathbf{j} + 1\mathbf{k}$$

$$\mathbf{N} \left(5, \tfrac{\pi}{4}, 0 \right) = \frac{5\sqrt{2}\mathbf{i} + 0\mathbf{j} + 5\sqrt{2}\mathbf{k}}{\sqrt{100}} \approx 0.7071\mathbf{i} + 0\mathbf{j} + 0.071\mathbf{k}$$

$$\mathbf{N} \left(5, \tfrac{\pi}{4}, \tfrac{\pi}{2} \right) = \frac{0\mathbf{i} + 5\sqrt{2}\mathbf{j} + 5\sqrt{2}\mathbf{k}}{\sqrt{100}} \approx 0\mathbf{i} + 0.7071\mathbf{j} + 0.071\mathbf{k}$$

as shown in Fig. 7.17. There is no unique tangent vector, only a unique tangent plane.

7.4.4 Unit Tangent and Normal Vectors to a Torus

Lastly, let's find the tangent and normal vectors for a torus. The equation for a torus with major radius R and minor radius r is

Fig. 7.17 Five unit normal
vectors for a sphere

$$\mathbf{r}(\theta, \phi) = \begin{bmatrix} (R + r \cos \theta) \cos \phi \\ (R + r \cos \theta) \sin \phi \\ r \sin \theta \end{bmatrix}, \quad (\theta, \phi) \in [0, 2\pi].$$

The tangent vectors are given by $\frac{\partial \mathbf{r}}{\partial \phi}$ and $\frac{\partial \mathbf{r}}{\partial \theta}$:

$$\frac{\partial \mathbf{r}}{\partial \phi} = \begin{bmatrix} -(R + r \cos \theta) \sin \phi \\ (R + r \cos \theta) \cos \phi \\ 0 \end{bmatrix}, \quad \frac{\partial \mathbf{r}}{\partial \theta} = \begin{bmatrix} -r \sin \theta \cos \phi \\ -r \sin \theta \sin \phi \\ r \cos \theta \end{bmatrix}.$$

For example, let $R = 3$ and $r = 1$, then:

$$\frac{\partial \mathbf{r}}{\partial \phi} = \begin{bmatrix} -(3 + \cos \theta) \sin \phi \\ (3 + \cos \theta) \cos \phi \\ 0 \end{bmatrix}, \quad \frac{\partial \mathbf{r}}{\partial \theta} = \begin{bmatrix} -\sin \theta \cos \phi \\ -\sin \theta \sin \phi \\ \cos \theta \end{bmatrix}.$$

When $\theta = \phi = 0$:

$$\frac{\partial \mathbf{r}}{\partial \phi} = \begin{bmatrix} 0 \\ 4 \\ 0 \end{bmatrix}, \quad \frac{\partial \mathbf{r}}{\partial \theta} = \begin{bmatrix} 0 \\ 0 \\ 1 \end{bmatrix}$$

which are shown in Fig. 7.18 as unit vectors.

If we now compute the cross product $\frac{\partial \mathbf{r}}{\partial \phi} \times \frac{\partial \mathbf{r}}{\partial \theta}$, we obtain the normal vector at that point:

$$\mathbf{N} = \begin{vmatrix} \mathbf{i} & \mathbf{j} & \mathbf{k} \\ 0 & 4 & 0 \\ 0 & 0 & 1 \end{vmatrix} = 4\mathbf{i} + 0\mathbf{j} + 0\mathbf{k}$$

which is shown in Fig. 7.18.

Fig. 7.18 The tangent vectors (green) and normal vector (red) on the left are for $\theta = \phi = 0$. The vectors on the right are for $\theta = \phi = \pi/2$

Let's compute a similar set of vectors for $\theta = \phi = \pi/2$:

$$\frac{\partial \mathbf{r}}{\partial \phi} = \begin{bmatrix} -\left(3 + \cos\left(\frac{\pi}{2}\right)\right)\sin\left(\frac{\pi}{2}\right) \\ \left(3 + \cos\left(\frac{\pi}{2}\right)\right)\cos\left(\frac{\pi}{2}\right) \\ 0 \end{bmatrix}, \qquad \frac{\partial \mathbf{r}}{\partial \theta} = \begin{bmatrix} -\sin\left(\frac{\pi}{2}\right)\cos\left(\frac{\pi}{2}\right) \\ -\sin\left(\frac{\pi}{2}\right)\sin\left(\frac{\pi}{2}\right) \\ \cos\theta \end{bmatrix}$$

$$\frac{\partial \mathbf{r}}{\partial \phi} = \begin{bmatrix} -3 \\ 0 \\ 0 \end{bmatrix}, \qquad \frac{\partial \mathbf{r}}{\partial \theta} = \begin{bmatrix} 0 \\ -1 \\ 0 \end{bmatrix}$$

which are shown in Fig. 7.18 as unit vectors.

If we now compute the cross product $\frac{\partial \mathbf{r}}{\partial \phi} \times \frac{\partial \mathbf{r}}{\partial \theta}$, we obtain the normal vector at that point:

$$\mathbf{N} = \begin{vmatrix} \mathbf{i} & \mathbf{j} & \mathbf{k} \\ -3 & 0 & 0 \\ 0 & -1 & 0 \end{vmatrix} = 0\mathbf{i} + 0\mathbf{j} + 3\mathbf{k}$$

which is shown in Fig. 7.18.

7.5 Summary

This chapter has shown how to calculate tangent and normal vectors to various curves and surfaces. The very same techniques can be applied to other curves and surfaces, but there is no guarantee that normalising vectors will always be an easy calculation.

7.5.1   Summary of Formulae

Unit tangent vector

$$\mathbf{r}(t) = x(t)\mathbf{i} + y(t)\mathbf{j} + z(t)\mathbf{k}$$
$$\mathbf{T}(t) = \frac{\mathbf{r}'(t)}{||\mathbf{r}'(t)||}.$$

Unit normal vector

$$\mathbf{N}(t) = \frac{\mathbf{T}'(t)}{||\mathbf{T}'(t)||}.$$

Unit tangent and normal vector to a line
Line in $\mathbb{R}^2$:

$$\mathbf{r}(t) = (x_1 + x_s t)\mathbf{i} + (y_1 + y_s t)\mathbf{j}$$
$$\mathbf{r}'(t) = x_s\mathbf{i} + y_s\mathbf{j}$$
$$\mathbf{T} = \frac{x_s\mathbf{i} + y_s\mathbf{j}}{\sqrt{x_s^2 + y_s^2}} = \lambda_1\mathbf{i} + \lambda_2\mathbf{j}$$
$$\mathbf{N} = -\lambda_2\mathbf{i} + \lambda_1\mathbf{j}, \quad \text{or} \quad = \lambda_2\mathbf{i} - \lambda_1\mathbf{j}.$$

Line in $\mathbb{R}^3$:

$$\mathbf{r}(t) = (x_1 + x_s t)\mathbf{i} + (y_1 + y_s t)\mathbf{j} + (z_1 + z_s t)\mathbf{k}$$
$$\mathbf{r}'(t) = x_s\mathbf{i} + y_s\mathbf{j} + z_s\mathbf{k}$$
$$\mathbf{T} = \frac{x_s\mathbf{i} + y_s\mathbf{j} + z_s\mathbf{k}}{\sqrt{x_s^2 + y_s^2 + z_s^2}}.$$

Unit tangent and normal vector to a circle

$$\mathbf{r}(t) = r\cos t\mathbf{i} + r\sin t\mathbf{j}$$
$$\mathbf{T}(t) = -\sin t\mathbf{i} + \cos t\mathbf{j}$$
$$\mathbf{N}(t) = -\cos t\mathbf{i} - \sin t\mathbf{j}.$$

Unit tangent and normal vector to an ellipse

$$\mathbf{r}(t) = a\cos t\mathbf{i} + b\sin t\mathbf{j}$$
$$\epsilon = \sqrt{1 - b^2/a^2}$$
$$\mathbf{T}(t) = \frac{-a\sin t\mathbf{i} + b\cos t\mathbf{j}}{a\sqrt{1 - \epsilon^2\cos^2 t}} = \lambda_1\mathbf{i} + \lambda_2\mathbf{j}$$
$$\mathbf{N} = -\lambda_2\mathbf{i} + \lambda_1\mathbf{j} \quad \text{or} \quad = \lambda_2\mathbf{i} - \lambda_1\mathbf{j}.$$

Unit tangent and normal vector to a quadratic Bézier curve

$$\mathbf{r}(t) = \mathbf{P}_0(1 - t)^2 + 2\mathbf{P}_1 t(1 - t) + \mathbf{P}_2 t^2, \quad t \in [0, 1]$$
$$x(t) = \mathbf{B}_{2,0}(t)x_0 + \mathbf{B}_{2,1}(t)x_1 + \mathbf{B}_{2,2}(t)x_2$$
$$y(t) = \mathbf{B}_{2,0}(t)y_0 + \mathbf{B}_{2,1}(t)y_1 + \mathbf{B}_{2,2}(t)y_2$$
$$x'(t) = 2[(x_1 - x_0)(1 - t) + (x_2 - x_1)t]$$
$$y'(t) = 2[(y_1 - y_0)(1 - t) + (y_2 - y_1)t]$$
$$\mathbf{T}(t) = \frac{2\big((\mathbf{P}_1 - \mathbf{P}_0)(1 - t) + (\mathbf{P}_2 - \mathbf{P}_1)t\big)}{\sqrt{\left(x'(t)\right)^2 + \left(y'(t)\right)^2}} = \lambda_1\mathbf{i} + \lambda_2\mathbf{j}$$
$$\mathbf{N} = -\lambda_2\mathbf{i} + \lambda_1\mathbf{j}, \quad \text{or} \quad = \lambda_2\mathbf{i} - \lambda_1\mathbf{j}.$$

Unit tangent and normal vector to a bilinear patch

$$L_1 = (\mathbf{P}_0, \mathbf{P}_1)$$
$$L_2 = (\mathbf{P}_2, \mathbf{P}_3)$$
$$\mathbf{r}(u, v) = (1 - v)(1 - u)\mathbf{P}_0 + u(1 - v)\mathbf{P}_1 + v(1 - u)\mathbf{P}_2 + uv\mathbf{P}_3$$
$$\frac{\partial \mathbf{r}}{\partial u} = (1 - v)(\mathbf{P}_1 - \mathbf{P}_0) + v(\mathbf{P}_3 - \mathbf{P}_2) = \lambda_{u1}\mathbf{i} + \lambda_{u2}\mathbf{j} + \lambda_{u3}\mathbf{k}$$
$$\frac{\partial \mathbf{r}}{\partial v} = (1 - u)(\mathbf{P}_2 - \mathbf{P}_0) + u(\mathbf{P}_3 - \mathbf{P}_1) = \lambda_{v1}\mathbf{i} + \lambda_{v2}\mathbf{j} + \lambda_{v3}\mathbf{k}$$
$$\mathbf{T}'(u, v) = \frac{\partial \mathbf{r}}{\partial u} \times \frac{\partial \mathbf{r}}{\partial v} = \begin{vmatrix} \mathbf{i} & \mathbf{j} & \mathbf{k} \\ \lambda_{u1} & \lambda_{u2} & \lambda_{u3} \\ \lambda_{v1} & \lambda_{v2} & \lambda_{v3} \end{vmatrix}$$
$$\mathbf{N}(t) = \frac{\mathbf{T}'(t)}{\|\mathbf{T}'(t)\|}.$$

Unit normal vector to a sphere

$$f(x, y, z) = x^2 + y^2 + z^2 - r^2$$
$$\nabla f = 2x\mathbf{i} + 2y\mathbf{j} + 2z\mathbf{k}$$
$$\mathbf{N}(x, y, z) = \frac{\nabla f}{\|\nabla f\|}$$
$$x = r \sin \phi \cos \theta$$
$$y = r \sin \phi \sin \theta$$
$$z = r \cos \phi$$
$$\nabla f(r, \phi, \theta) = 2r \sin \phi \cos \theta\mathbf{i} + 2r \sin \phi \sin \theta\mathbf{j} + 2r \cos \phi\mathbf{k}.$$

Unit tangent and normal vector to a torus

$$\mathbf{r}(\theta, \phi) = \begin{bmatrix} (R + r\cos\theta)\cos\phi \\ (R + r\cos\theta)\sin\phi \\ r\sin\theta \end{bmatrix}, \quad (\theta, \phi) \in [0, 2\pi]$$

$$\frac{\partial \mathbf{r}}{\partial \phi} = \begin{bmatrix} -(R + r\cos\theta)\sin\phi \\ (R + r\cos\theta)\cos\phi \\ 0 \end{bmatrix}, \quad \frac{\partial \mathbf{r}}{\partial \theta} = \begin{bmatrix} -r\sin\theta\cos\phi \\ -r\sin\theta\sin\phi \\ r\cos\theta \end{bmatrix}$$

$$\mathbf{N}(\theta, \phi) = \frac{\partial \mathbf{r}}{\partial \phi} \times \frac{\partial \mathbf{r}}{\partial \theta}.$$

Chapter 8
Straight Lines

8.1 Introduction

This chapter explores different ways of using vectors to represent straight lines and line segments. We begin with parametric line equations, followed by the Cartesian form, and the general form of the line equation. Then we examine 2-D space partitioning, where we exploit the ability of line equations to partition space. We then introduce perpendicular vectors and show how they are used to resolve geometric problems such as finding a line perpendicular to a vector, and the position of a point reflected in a line. Finally, we examine how to use vectors to represent line segments, and the intersection of two straight lines, and two line segments.

8.2 Line Equations

Having explored the basics of vector analysis, let us apply some of these ideas to the description of straight lines. There are various ways of describing straight lines that include:

- The normal form of the line equation $y = mx + c$.
- The general form of the line equation $ax + by + c = 0$.
- The Cartesian form of the line equation $ax + by = c$.
- The parametric form of the line equation $\mathbf{p} = \mathbf{t} + \lambda \hat{\mathbf{v}}$.

The first three only apply to 2-D straight lines, whereas the fourth applies to 2-D and 3-D straight lines. So let's begin with this.

© Springer-Verlag London Ltd., part of Springer Nature 2021
J. Vince, *Vector Analysis for Computer Graphics*,
https://doi.org/10.1007/978-1-4471-7505-6_8

8.2.1   The Parametric Form of the Line Equation

The first objective is to find a way to move along a straight line using a scalar parameter. Fortunately, vectors provide such a mechanism in the form of the operation $\lambda \hat{\mathbf{v}}$, which scales the unit vector $\hat{\mathbf{v}}$ with the scalar λ. But we also need to relate our position relative to some reference point, which must be included in our definition. We begin as follows.

Figure 8.1 shows a line passing through the point $T(x_t, y_t)$, where $P(x_p, y_p)$ is a point on the line. The unit vector $\hat{\mathbf{v}}$ provides the line's direction. From this scenario we can state

$$\overrightarrow{OP} = \overrightarrow{OT} + \overrightarrow{TP}.$$

Given:

- $\hat{\mathbf{v}}$ is a unit vector defining the line's direction,
- λ is a parameter to position us anywhere along the direction of $\hat{\mathbf{v}}$,
- $T(x_t, y_t)$ is a reference point on the line,
- $P(x_p, y_p)$ is any point on the line,
- $\mathbf{t}$ is T's position vector,
- $\mathbf{p}$ is P's position vector.

Then,

$$\mathbf{p} = \mathbf{t} + \lambda \hat{\mathbf{v}}. \tag{8.1}$$

It should be clear that (8.1) works equally well in 3-D as it does in 2-D, so let us bring it alive with the following conditions:

Fig. 8.1 A 2-D parametric line

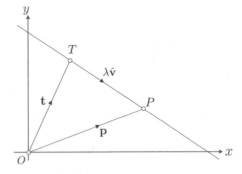

Fig. 8.2 A 3-D parametric line

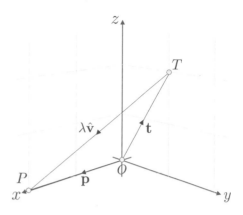

$$T = (1, 2)$$
$$\mathbf{t} = 1\mathbf{i} + 2\mathbf{j}$$
$$\mathbf{v} = 1\mathbf{i} - 1\mathbf{j}$$
$$\hat{\mathbf{v}} = \tfrac{1}{\sqrt{2}}(1\mathbf{i} - 1\mathbf{j})$$
$$\mathbf{p} = 1\mathbf{i} + 2\mathbf{j} + \tfrac{\lambda}{\sqrt{2}}(1\mathbf{i} - 1\mathbf{j}) = \left(1 + \tfrac{\lambda}{\sqrt{2}}\right)\mathbf{i} + \left(2 - \tfrac{\lambda}{\sqrt{2}}\right)\mathbf{j}.$$

When $\lambda = 0$, $\mathbf{p} = 1\mathbf{i} + 2\mathbf{j}$, and when $\lambda = 1$, $\mathbf{p} = (1 + \tfrac{1}{\sqrt{2}})\mathbf{i} + (2 - \tfrac{1}{\sqrt{2}})\mathbf{j} \approx (1.707, 1.293)$. Thus $(1.707, 1.293)$ is a point one unit along $\hat{\mathbf{v}}$ from T. By changing the value and sign of λ one can move forward or backwards along $\hat{\mathbf{v}}$.

Let's illustrate (8.1) in a 3-D context as shown in Fig. 8.2. Let

$$T = (0, 1, 2)$$
$$\mathbf{t} = 0\mathbf{i} + 1\mathbf{j} + 2\mathbf{k}$$
$$\mathbf{v} = 2\mathbf{i} - 1\mathbf{j} - 2\mathbf{k}$$
$$\hat{\mathbf{v}} = \tfrac{1}{3}(2\mathbf{i} - 1\mathbf{j} - 2\mathbf{k})$$
$$\mathbf{p} = 0\mathbf{i} + 1\mathbf{j} + 2\mathbf{k} + \tfrac{\lambda}{3}(2\mathbf{i} - 1\mathbf{j} - 2\mathbf{k}) = \tfrac{2\lambda}{3}\mathbf{i} + \left(1 - \tfrac{\lambda}{3}\right)\mathbf{j} + \left(2 - \tfrac{2\lambda}{3}\right)\mathbf{k}.$$

When $\lambda = 0$, $\mathbf{p} = 0\mathbf{i} + 1\mathbf{j} + 2\mathbf{k}$, and when $\lambda = 1$, $\mathbf{p} = \tfrac{2}{3}\mathbf{i} + \tfrac{2}{3}\mathbf{j} + \tfrac{4}{3}\mathbf{k} \approx (0.667, 0.667, 1.333)$. Thus $(0.667, 0.667, 1.333)$ is a point one unit along $\hat{\mathbf{v}}$ from T. Similarly, by changing the value of λ, we can move forward and backward along the 3-D line.

As we may not always have an explicit description of $\hat{\mathbf{v}}$, let's consider the case when the line is represented by two points in space. If the two points are $P_1(x_1, y_1)$ and $P_2(x_2, y_2)$ we can state that

$$\mathbf{v} = (x_2 - x_1)\mathbf{i} + (y_2 - y_1)\mathbf{j}$$

from which we can compute $\|\mathbf{v}\|$. $\hat{\mathbf{v}}$ is then equal to $\tfrac{\mathbf{v}}{\|\mathbf{v}\|}$ and T can be P_1.

Fig. 8.3 A 2-D parametric
line

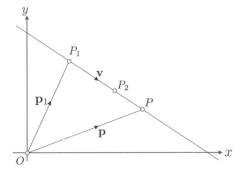

We can now write (8.1) as follows:

$$\mathbf{p} = \mathbf{p}_1 + \lambda \hat{\mathbf{v}} \tag{8.2}$$

where $\mathbf{p}_1$ is the position vector for P_1. This scenario is shown in Fig. 8.3.

In the current $\mathbb{R}^2$ context (8.2) reveals:

$$x = x_1 + \lambda \hat{\mathbf{v}}$$

$$y = y_1 + \lambda \hat{\mathbf{v}}$$

$$x = x_1 + \frac{\lambda}{\|\mathbf{v}\|}(x_2 - x_1)$$

$$y = y_1 + \frac{\lambda}{\|\mathbf{v}\|}(y_2 - y_1)$$

$$x = \left(1 - \frac{\lambda}{\|\mathbf{v}\|}\right)x_1 + \frac{\lambda}{\|\mathbf{v}\|}x_2 \tag{8.3}$$

$$y = \left(1 - \frac{\lambda}{\|\mathbf{v}\|}\right)y_1 + \frac{\lambda}{\|\mathbf{v}\|}y_2. \tag{8.4}$$

It is clear from (8.3) and (8.4) that when $\lambda = \|\mathbf{v}\|$, $x = x_2$, and $y = y_2$, and represents a point (x, y), $\|\mathbf{v}\|$ units along the line from P_1.

Let us follow this parametric description of a straight line with the derivation of the Cartesian form of the line equation.

8.2.2   The Cartesian Form of the Line Equation

We know that the Cartesian form of the line equation is $ax + by = c$, but what is not obvious is the geometric significance of a, b and c. Fortunately, vector analysis provides an insight into these terms, which is the objective of this section.

Fig. 8.4 Step 1

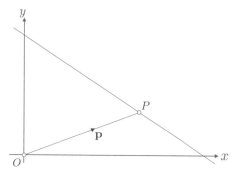

Fig. 8.5 Step 2

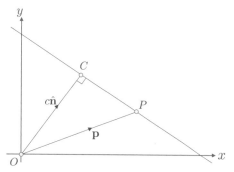

We begin by setting out to discover the coordinates of any point $P(x, y)$ on a 2-D straight line. The strategy used may not immediately appear intuitive, but try following it, and hopefully the approach will become clear.

Step 1

Construct a line away from the origin O containing the point $P(x, y)$ and its associated position vector **p** as shown in Fig. 8.4.

We now need to construct a second path from O to P.

Step 2

Construct a line from O to a point C on the line such that $\overrightarrow{OC}$ is perpendicular to the line. It is convenient to make $\overrightarrow{OC}$ a scalar multiple of a unit vector $\hat{\mathbf{n}} = a\mathbf{i} + b\mathbf{j}$. i.e. $\overrightarrow{OC} = c\hat{\mathbf{n}}$, where c is some scalar as shown in Fig. 8.5.

Step 3

Extend the diagram to include the vector $\overrightarrow{CP}$, which completes the second path from O to P, as shown in Fig. 8.6. We can now write $\overrightarrow{OP} = \overrightarrow{OC} + \overrightarrow{CP}$. Substituting vector names:

$$\mathbf{p} = c\hat{\mathbf{n}} + \overrightarrow{CP}. \tag{8.5}$$

Fig. 8.6 Step 3

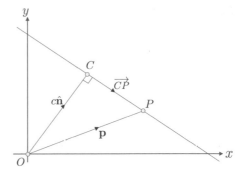

Although we know c and $\hat{\mathbf{n}}$, and the fact that $\mathbf{p}$ points to any point on the line, we do not know $\overrightarrow{CP}$. Somehow, $\overrightarrow{CP}$ has to be eliminated, which can be achieved by the following cunning subterfuge.

The dot product of two perpendicular vectors is zero, therefore $c\hat{\mathbf{n}} \cdot \overrightarrow{CP} = 0$. In fact, the scalar c is superfluous, therefore $\hat{\mathbf{n}} \cdot \overrightarrow{CP} = 0$. Unfortunately (8.5) does not contain such a term, but there is nothing to stop us from introducing one by multiplying (8.5) by $\hat{\mathbf{n}}$ using the dot product:

$$\hat{\mathbf{n}} \cdot \mathbf{p} = c\hat{\mathbf{n}} \cdot \hat{\mathbf{n}} + \hat{\mathbf{n}} \cdot \overrightarrow{CP}$$

which reduces to:

$$\hat{\mathbf{n}} \cdot \mathbf{p} = c\hat{\mathbf{n}} \cdot \hat{\mathbf{n}}. \tag{8.6}$$

But $\hat{\mathbf{n}} \cdot \hat{\mathbf{n}} = 1$, we can write (8.6) as

$$\hat{\mathbf{n}} \cdot \mathbf{p} = c \tag{8.7}$$

which expands to:

$$(a\mathbf{i} + b\mathbf{j}) \cdot (x\mathbf{i} + y\mathbf{j}) = c$$
$$ax + by = c.$$

Which is the Cartesian form of the line equation where:

- x and y are the coordinates of a point on the line,
- a and b are the components of a unit vector normal to the line,
- c is the perpendicular distance from the origin to the line.

Although the above interpretation is correct for our example, we must be careful not to be too hasty when interpreting straight line equations. For example, in (8.8),

$$3x + 4y = 10 \tag{8.8}$$

Fig. 8.7 The graph of $0.6x + 0.8y = 2$

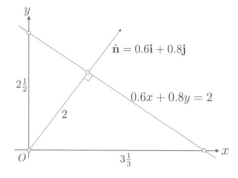

Fig. 8.8 The graph of $-0.6x - 0.8y = 2$

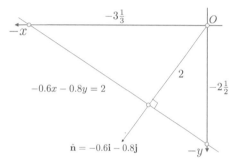

10 does not represent the perpendicular distance from the origin to the line. This is because 3 and 4 are not the components of a unit vector—they are the components of a vector 5 units long.

Therefore, if we divide (8.8) by 5 we obtain:

$$0.6x + 0.8y = 2 \qquad (8.9)$$

where 0.6 and 0.8 are the components of a unit vector, and 2 is the perpendicular distance from the origin to the line. Figure 8.7 shows the graph of the line Eq. (8.9).

So far, all this seems reasonable and consistent, but how do we interpret (8.10) where c is negative?

$$0.6x + 0.8y = -2. \qquad (8.10)$$

Clearly, the Euclidean perpendicular distance from the origin to the line cannot be negative. Furthermore, Fig. 8.8 shows that the normal vector $\hat{\mathbf{n}} = -0.6\mathbf{i} - 0.8\mathbf{j}$, which provides the real clue. These two facts are simply reminding us to rewrite (8.10) in a form that preserves a positive c term, which is achieved by multiplying (8.10) by -1:

$$-0.6x - 0.8y = 2. \qquad (8.11)$$

Now the vector components are correct and the perpendicular distance from the origin to the line is $+2$ (8.11), with a positive c term makes geometric sense, and makes the normal vector components explicit.

8.2.3   The General Form of the Line Equation

The general form of the line equation is:

$$ax + by + c = 0 \tag{8.12}$$

and to convert it to its equivalent Cartesian form we require two things: the c scalar must be negative, and a and b must represent a unit vector.

For example, given (8.13)

$$5x + 12y + 26 = 0. \tag{8.13}$$

We convert the 26 into -26 by multiplying throughout by -1:

$$-5x - 12y - 26 = 0. \tag{8.14}$$

The magnitude of the perpendicular vector is $\sqrt{(-5)^2 + (-12)^2} = 13$, so we divide throughout by 13:

$$-\tfrac{5}{13}x - \tfrac{12}{13}y - 2. \tag{8.15}$$

Finally, we write (8.15) in its Cartesian form:

$$-\tfrac{5}{13}x - \tfrac{12}{13}y = 2 \tag{8.16}$$

where the unit perpendicular vector is $\hat{\mathbf{n}} = -\tfrac{5}{13}\mathbf{i} - \tfrac{12}{13}\mathbf{j}$, and the line is 2 units from the origin.

8.3   2-D Space Partitioning

One recurring problem in computer graphics is whether a point is inside or outside a 2-D area or 3-D volume. There are various solutions to this problem, but let's see how vector analysis provides some assistance.

Equation (8.17) shows the general form of a line equation, whose graph is shown in Fig. 8.7.

$$0.6x + 0.8y - 2 = 0. \tag{8.17}$$

When we substitute values of (x, y) for points on the line, the expression (8.18)

$$0.6x + 0.8y - 2 \qquad (8.18)$$

equals zero. But what happens if the point is off the line? Well, let's try substituting some points.

The point $(10, 10)$ is clearly above the line, and when substituted in (8.18) we obtain the value $+12$. However, the point $(-10, -10)$ is below the line, and produces the value -16.

If we continued substituting other points in (8.18), we would discover that the line partitions space into three zones: a zone above the line where (8.18) is positive, and a zone below the line where (8.18) is negative, and points on the line where (8.18) is zero. But the terms 'above' and 'below' are incorrect, as they have no meaning for vertical lines!

To discover the key to this problem we need to see what happens when the line's normal vector is reversed:

$$-0.6x - 0.8y - 2 = 0. \qquad (8.19)$$

whose graph is shown in Fig. 8.8. This time the expression is

$$-0.6x - 0.8y - 2. \qquad (8.20)$$

Substituting the point $(10, 10)$ in (8.20) equals -16, whereas $(-10, -10)$ equals $+12$. Notice that for both equations, the positive sign is in the zone containing the normal vector, which is always the case. The reader may wish to reason why this is so.

Therefore, if we had a convex boundary formed by straight edges, how can we arrange that their normal vectors point outwards or inwards? This is what we'll investigate next. Ideally, we want to create one of the scenarios shown in Fig. 8.9.

We begin by making the convex boundary from a list of vertices P_1, P_2, P_3 and P_4, defined in a counter-clockwise sequence, which create a corresponding chain of

Fig. 8.9 Normal vectors pointing inwards and outwards

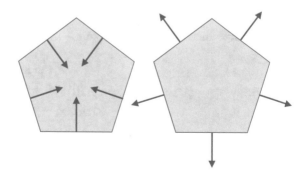

Fig. 8.10 A quadrilateral
composed of four vectors

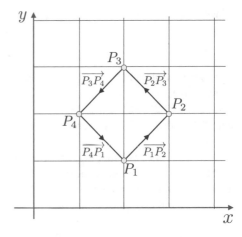

Fig. 8.11 One side of the
quadrilateral

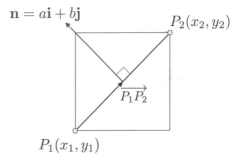

vectors. For instance, a quadrilateral has been chosen as shown in Fig. 8.10. We now
need to find the general form of the line equation for each boundary vector. If we do
not apply a consistent algorithm, we could end up with an incoherent collection of
normal vectors.

We start with two points $P_1(x_1, y_1)$ and $P_2(x_2, y_2)$ that are the ends of a directed
line, as shown in Fig. 8.11. Next we let

$$\mathbf{n} = a\mathbf{i} + b\mathbf{j}$$

which is perpendicular to $\overrightarrow{P_1P_2}$. Therefore,

$$\overrightarrow{P_1P_2} \cdot \mathbf{n} = 0$$
$$[(x_2 - x_1)\mathbf{i} \quad (y_2 - y_1)\mathbf{j}] \cdot [a\mathbf{i} \quad b\mathbf{j}] = 0$$
$$a(x_2 - x_1) + b(y_2 - y_1) = 0$$
$$\frac{a}{b} = \frac{y_1 - y_2}{x_2 - x_1}$$

and

$$a = y_1 - y_2, \quad b = x_2 - x_1.$$

The general form of the line equation is:

$$(y_1 - y_2)x + (x_2 - x_1)y + c = 0 \qquad (8.21)$$

and c is found by substituting P_1 or P_2 in (8.21).

We now evaluate (8.21) for each boundary edge.

$\overrightarrow{P_1P_2}$: Substituting $x_1 = 2$, $y_1 = 1$, $x_2 = 3$, $y_2 = 2$ in (8.21) we get

$$-x + y + c = 0 \qquad (8.22)$$

and substituting the point $P_1(2, 1)$ in (8.22) makes $c = 1$.
The general form of the line equation is:

$$-x + y + 1 = 0$$

with a normal vector:

$$\mathbf{n}_1 = -\mathbf{i} + \mathbf{j}.$$

$\overrightarrow{P_2P_3}$: Substituting $x_1 = 3$, $y_1 = 2$, $x_2 = 2$, $y_2 = 3$ in (8.21) we get

$$-x - y + c = 0 \qquad (8.23)$$

and substituting the point $P_2(3, 2)$ in (8.23) makes $c = 5$.
The general form of the line equation is:

$$-x - y + 5 = 0$$

with a normal vector:

$$\mathbf{n}_2 = -\mathbf{i} - \mathbf{j}.$$

$\overrightarrow{P_3P_4}$: Substituting $x_1 = 2$, $y_1 = 3$, $x_2 = 1$, $y_2 = 2$ in (8.21) we get

$$x - y + c = 0 \qquad (8.24)$$

and substituting the point $P_3(2, 3)$ in (8.24) makes $c = 1$.
The general form of the line equation is:

$$x - y + 1 = 0$$

with a normal vector:

$$\mathbf{n}_3 = \mathbf{i} - \mathbf{j}.$$

Fig. 8.12 The scaled four
normal vectors

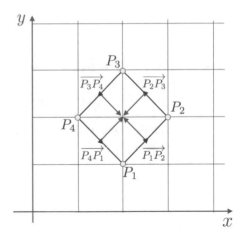

$\overrightarrow{P_4P_1}$: Substituting $x_1 = 1, y_1 = 2, x_2 = 2, y_2 = 1$ in (8.21) we get

$$x + y + c = 0 \qquad\qquad (8.25)$$

and substituting the point $P_4(1, 2)$ in (8.25) makes $c = -3$.
The general form of the line equation is:

$$x + y - 3 = 0$$

with a normal vector:

$$\mathbf{n}_4 = \mathbf{i} + \mathbf{j}.$$

Figure 8.12 shows the boundary with the four normal vectors superimposed, and
we see that they all point to the boundary's inside. The four expressions can now be
tested with different points, and it is the sign of the expression that confirms their
space partition.

If only one expression is zero, the point is on the boundary.
If two expressions are zero, the point is a boundary vertex.
If any expression is negative, the point is outside the boundary.
If all expressions are positive, the point is inside the boundary.

Table 8.1 summarises some of these conditions. The reader may wish to verify other
points, but the contents of Table 8.1 should be sufficient to demonstrate the integrity of
this strategy. However, this technique is sensitive to the boundary's vertex sequence.
For if the vertices are taken in a clockwise order, the four line equations are effectively
multiplied by -1, which flips the normal vectors:

Table 8.1 Partition for various points

Point	$-x + y + 1$	$-x - y + 5$	$x - y + 1$	$x + y - 3$	Partition
(2, 2)	+	+	+	+	Inside
(1, 1)	+	+	+	−	Outside
(2, 0)	−	+	+	−	Outside
(3, 1)	−	+	+	+	Outside
(4, 2)	−	−	+	+	Outside
(3, 3)	+	−	+	+	Outside
(2, 4)	+	−	−	+	Outside
(1, 3)	+	+	−	+	Outside
(0, 2)	+	+	−	−	Outside
(2, 1)	0	+	+	0	Vertex
(3, 2)	0	0	+	+	Vertex
(2, 3)	+	0	0	+	Vertex
(1, 2)	+	+	0	0	Vertex

$$-x + y + 1 \Rightarrow x - y - 1$$
$$-x - y + 5 \Rightarrow x + y - 5$$
$$x - y + 1 \Rightarrow -x + y - 1$$
$$x + y - 3 \Rightarrow -x - y + 3.$$

Now the normal vectors point away from the boundary's inside, which makes the inside partition negative and the outside positive. This still works, but highlights the care required when using this technique.

8.4   Perpendicular Vectors

When working in 2-D we often employ vectors that are perpendicular to some reference vector. For instance, Fig. 8.13 shows two vectors **v** and **n**, where **n** is perpendicular to **v**, and is expressed mathematically as **n** ⊥ **v**.

Fig. 8.13 Two perpendicular vectors

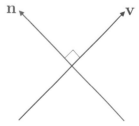

Fig. 8.14 Two arrangements of perpendicular vectors

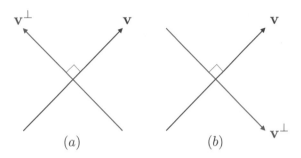

Fig. 8.15 Vectors **v** and $\mathbf{v}^{\perp}$

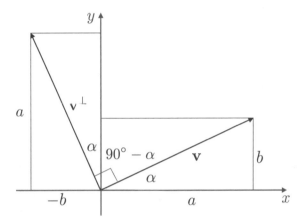

The symbol '$\perp$', (pronounced 'perp') can be regarded as an operator, such that given a vector **v**, $\mathbf{v}^{\perp}$ is a vector perpendicular to **v**. However, there is a problem: which way does $\mathbf{v}^{\perp}$ point? Figure 8.14 shows the two possibilities. Either one is valid, however mathematics does employ a convention where a counter-clockwise rotation is positive. Consequently, the orientation shown in Fig. 8.14a is the one adopted. The next step is to find the components of $\mathbf{v}^{\perp}$.

Figure 8.15 shows vector $\mathbf{v} = a\mathbf{i} + b\mathbf{j}$ which makes an angle α with the x-axis, and $90° - \alpha$ with the y-axis. If we transpose **v**'s components to $-b\mathbf{i} + a\mathbf{j}$ we create a second vector, which must be perpendicular to **v**, as the angle between the two vectors is $90°$. Therefore, we can state

$$\mathbf{v} = a\mathbf{i} + b\mathbf{j}$$
$$\mathbf{v}^{\perp} = -b\mathbf{i} + a\mathbf{j}. \qquad (8.26)$$

Further confirmation is found in the dot product where

$$\mathbf{v} \cdot \mathbf{v}^{\perp} = (a\mathbf{i} + b\mathbf{j}) \cdot (-b\mathbf{i} + a\mathbf{j}) = -ab + ab = 0.$$

A second way of demonstrating this is to convert $\mathbf{v}$ into a complex number:

$$\mathbf{v} = a\mathbf{i} + b\mathbf{j} \equiv a + ib \qquad (8.27)$$

where,

$$i^2 = -1.$$

If we multiply (8.27) by i, it rotates it through 90°:

$$i(a + ib) = ai + i^2b = -b + ai$$

where,

$$-b + ai \equiv -b\mathbf{i} + a\mathbf{j}.$$

Although the sign change and component switching is a simple operation, it can be represented formally by this determinant:

$$\mathbf{v}^{\perp} = - \begin{vmatrix} \mathbf{i} & \mathbf{j} \\ a & b \end{vmatrix} = -b\mathbf{i} + a\mathbf{j}.$$

It may be obvious that the magnitude $\|\mathbf{v}^{\perp}\|$ equals the magnitude $\|\mathbf{v}\|$.

8.5 A Line Perpendicular to a Vector

In this section we examine how to define a straight line perpendicular to a given vector, and passes through a specific point. Figure 8.16 shows this scenario where the reference unit vector $\hat{\mathbf{n}}$ passes through the origin O and the specified point is $T(x_t, y_t)$. Figure 8.16 shows the following conditions:

Fig. 8.16 A line perpendicular to a vector

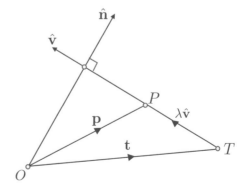

Fig. 8.17 A line
perpendicular to a vector

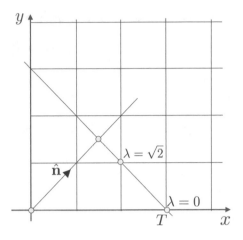

Fig. 8.17 A line
perpendicular to a vector

- $\hat{\mathbf{n}} = n_x\mathbf{i} + n_y\mathbf{j}$ is a unit vector,
- $\hat{\mathbf{v}} = -y_n\mathbf{i} + x_n\mathbf{j}$ is perpendicular to $\hat{\mathbf{n}}$, and directs the line,
- $T(x_t, y_t)$ is the reference point on the line,
- $\mathbf{t} = x_t\mathbf{i} + y_t\mathbf{j}$ is T's position vector,
- $P(x_p, y_p)$ is any point on the line,
- $\mathbf{p} = x_p\mathbf{i} + y_p\mathbf{j}$ is P's position vector,
- λ is a scalar and locates P relative to T.

From Fig. 8.16 we can state

$$\mathbf{p} = \mathbf{t} + \lambda(-y_n\mathbf{i} + x_n\mathbf{j}). \tag{8.28}$$

Therefore, from (8.28):

$$x_p = x_t - \lambda y_n \tag{8.29}$$
$$y_p = y_t + \lambda x_n. \tag{8.30}$$

With reference to Fig. 8.17, $T = (3, 0)$ and $\hat{\mathbf{n}} = \frac{1}{\sqrt{2}}(\mathbf{i} + \mathbf{j})$, therefore, using (8.29) and (8.30) we have:

When $\lambda = 0$, $P = (3, 0)$, and when $\lambda = \sqrt{2}$, $P = (2, 1)$.

8.6   The Position of a Point Reflected in a Line

In this section we consider the problem of calculating the reflection of a point in a straight line, and once more we consider two separate strategies using two forms of the line equation. In both cases, the key to the solution is based on the fact that the line connecting the point to its reflection is perpendicular to the reflecting line.

8.6.1 The Cartesian Form of the Line Equation

We start with the following conditions shown in Fig. 8.18:

- $ax + by = c$ is the line's equation,
- $P(x_p, y_p)$ is the point to be reflected,
- $Q(x_q, y_q)$ is P's reflection,
- $T(x_t, y_t)$ is a point on the line such that $\overrightarrow{TP}$ is orthogonal to the line,
- $\mathbf{p} = x_p\mathbf{i} + y_p\mathbf{j}$ is P's position vector,
- $\mathbf{q} = x_q\mathbf{i} + y_q\mathbf{j}$ is Q's position vector,
- $\mathbf{t} = x_t\mathbf{i} + y_t\mathbf{j}$ is T's position vector,
- $\mathbf{n} = a\mathbf{i} + b\mathbf{j}$ is the line's normal vector,
- $\lambda\mathbf{n}$ is the vector $\overrightarrow{TP}$ and $\overrightarrow{QT}$.

The objective is to produce a vector equation of the form $\mathbf{q} = \mathbf{p} + \cdots$. The RHS of this equation must only contain references to $\mathbf{p}$ and variables in the line equation. λ is acceptable, so long as it can be defined. As $\overrightarrow{QP}$ is orthogonal to the reflecting line, we must anticipate employing the dot product.

With reference to Fig. 8.18:

$$\mathbf{n} \cdot \mathbf{t} = (a\mathbf{i} + b\mathbf{j}) \cdot (x_t\mathbf{i} + y_t\mathbf{j}) = ax_t + by_t = c \tag{8.31}$$

and

$$\mathbf{p} = \mathbf{t} + \lambda\mathbf{n}. \tag{8.32}$$

Multiply (8.32) by $\mathbf{n}$ using the dot product,

$$\mathbf{n} \cdot \mathbf{p} = \mathbf{n} \cdot \mathbf{t} + \lambda\mathbf{n} \cdot \mathbf{n}. \tag{8.33}$$

Fig. 8.18 The reflection of a point P in a line

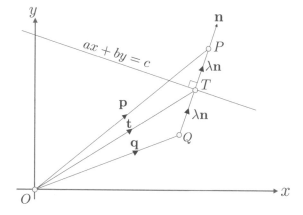

Substitute (8.31) in (8.33)

$$\mathbf{n} \cdot \mathbf{p} = c + \lambda \mathbf{n} \cdot \mathbf{n}. \tag{8.34}$$

Therefore, we have

$$\lambda = \frac{\mathbf{n} \cdot \mathbf{p} - c}{\mathbf{n} \cdot \mathbf{n}} = \frac{\mathbf{n} \cdot \mathbf{p} - c}{\|\mathbf{n}\|^2}. \tag{8.35}$$

Now,

$$\mathbf{q} = \mathbf{p} - 2\lambda \mathbf{n}. \tag{8.36}$$

Therefore,

$$\mathbf{q} = \mathbf{p} - 2\left(\frac{\mathbf{n} \cdot \mathbf{p} - c}{\|\mathbf{n}\|^2} \right) \mathbf{n}. \tag{8.37}$$

If $\mathbf{n}$ is a unit vector $\hat{\mathbf{n}}$, then

$$\mathbf{q} = \mathbf{p} - 2(\hat{\mathbf{n}} \cdot \mathbf{p} - c)\hat{\mathbf{n}}. \tag{8.38}$$

Let's test (8.37) with a simple example—one where we can predict the result.

Figure 8.19 shows the graph of the line equation $x + y = 3$. The reference point is $P(2, 0)$ and it is obvious that the reflected point must be $Q(3, 1)$. The conditions are:

- $x + y = 3$ is the Cartesian line equation,
- $\mathbf{n} = 1\mathbf{i} + 1\mathbf{j}$ is the normal to the line,
- $\|\mathbf{n}\| = \sqrt{2}$ is the magnitude of $\mathbf{n}$,
- $c = 3$ is from the line equation,
- $\mathbf{p} = 2\mathbf{i} + 0\mathbf{j}$ is P's position vector,
- $\mathbf{q} = x_q\mathbf{i} + y_q\mathbf{j}$ is Q position vector.

Fig. 8.19 The reflection of a point P in a line

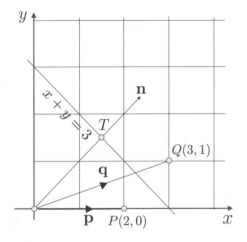

Fig. 8.20 The reflection of a
point P about a line

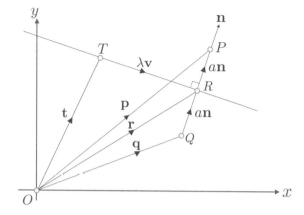

Therefore, using (8.37):

$$\mathbf{q} = (2\mathbf{i} + 0\mathbf{j}) - 2\left(\frac{(1\mathbf{i} + 1\mathbf{j}) \cdot (2\mathbf{i} + 0\mathbf{j}) - 3}{2}\right)(1\mathbf{i} + 1\mathbf{j})$$

$$= (2\mathbf{i} + 0\mathbf{j}) - (-1)(1\mathbf{i} + 1\mathbf{j})$$

$$= (2\mathbf{i} + 0\mathbf{j}) + (1\mathbf{i} + 1\mathbf{j})$$

$$= 3\mathbf{i} + 1\mathbf{j}$$

$$x_q = 3$$

$$y_q = 1.$$

As predicted.

8.6.2 *The Parametric Form of the Line Equation*

Now let's use the parametric form of the line equation. Figure 8.20 shows the following conditions:

- **v** is the line's direction vector,
- T is some point on the line,
- **t** is T's position vector,
- P is the point to be reflected,
- **p** is P's position vector,
- Q is P's reflection,
- **q** is Q's position vector,
- **n** is orthogonal to **v**,
- R is midway between P and Q,
- $\lambda\mathbf{v}$ is $\overrightarrow{TR}$.

The following analysis exploits the fact that $\overrightarrow{QP}$ is perpendicular to $\mathbf{v}$, and the objective is to find $\mathbf{q}$ in terms of $\mathbf{t}$, $\mathbf{p}$ and $\mathbf{v}$. We begin with

$$\mathbf{r} = \mathbf{t} + \lambda \mathbf{v}. \tag{8.39}$$

Multiply (8.39) by $\mathbf{v}$ using the dot product:

$$\mathbf{v} \cdot \mathbf{r} = \mathbf{v} \cdot \mathbf{t} + \lambda \mathbf{v} \cdot \mathbf{v}. \tag{8.40}$$

As $\mathbf{p}$ and $\mathbf{r}$ have a common projection on $\mathbf{v}$, we have

$$\mathbf{v} \cdot \mathbf{p} = \mathbf{v} \cdot \mathbf{r}. \tag{8.41}$$

Substitute (8.41) in (8.40):

$$\mathbf{v} \cdot \mathbf{p} = \mathbf{v} \cdot \mathbf{t} + \lambda \mathbf{v} \cdot \mathbf{v}. \tag{8.42}$$

Therefore,

$$\lambda = \frac{\mathbf{v} \cdot (\mathbf{p} - \mathbf{t})}{\mathbf{v} \cdot \mathbf{v}} = \frac{\mathbf{v} \cdot (\mathbf{p} - \mathbf{t})}{\|\mathbf{v}\|^2}. \tag{8.43}$$

If $\mathbf{v}$ is a unit vector $\hat{\mathbf{v}}$, then

$$\lambda = \hat{\mathbf{v}} \cdot (\mathbf{p} - \mathbf{t}). \tag{8.44}$$

From Fig. 8.20 we have

$$\mathbf{p} = \mathbf{r} + a\mathbf{n}$$
$$2\mathbf{p} = 2\mathbf{r} + 2a\mathbf{n}$$
$$\mathbf{q} = \mathbf{p} - 2a\mathbf{n}$$
$$\mathbf{q} + 2\mathbf{p} = \mathbf{p} + 2\mathbf{r}$$
$$\mathbf{q} = 2\mathbf{r} - \mathbf{p}. \tag{8.45}$$

Substitute (8.39) in (8.45):

$$\mathbf{q} = 2(\mathbf{t} + \lambda \mathbf{v}) - \mathbf{p}. \tag{8.46}$$

Let's test (8.46) with the previous example.
Figure 8.21 shows the following conditions:

- $\mathbf{v} = 1\mathbf{i} - 1\mathbf{j}$ is the line's direction vector,
- $T(1, 2)$ is on the line,
- $\mathbf{t} = 1\mathbf{i} + 2\mathbf{j}$ is T's position vector,
- $P(2, 0)$ is the point to be reflected,
- $\mathbf{p} = 2\mathbf{i} + 0\mathbf{j}$ is P's position vector,
- $Q(3, 1)$ is P's reflection,
- $\mathbf{q} = 3\mathbf{i} + 1\mathbf{j}$ is Q's position vector.

Fig. 8.21 The reflection of a
point P in a line

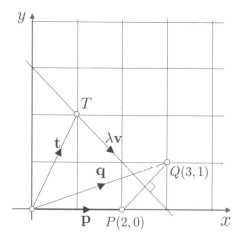

Using (8.43) and (8.46):

$$\lambda = \frac{\mathbf{v} \cdot (\mathbf{p} - \mathbf{t})}{\|\mathbf{v}\|^2}$$

$$= \frac{[1\mathbf{i} \quad - 1\mathbf{j}] \cdot [[2\mathbf{i} \quad 0\mathbf{j}] - [1\mathbf{i} \quad 2\mathbf{j}]]}{2}$$

$$= \frac{[1\mathbf{i} \quad - 1\mathbf{j}] \cdot [1\mathbf{i} \quad - 2\mathbf{j}]}{2} = 1.5$$

$$\mathbf{q} = 2(\mathbf{t} + \lambda\mathbf{v}) - \mathbf{p}$$

$$= 2\big((1\mathbf{i} + 2\mathbf{j}) + 1.5(1\mathbf{i} - 1\mathbf{j})\big) - (2\mathbf{i} + 0\mathbf{j})$$

$$= 2(2.5\mathbf{i} + 0.5\mathbf{j}) - (2\mathbf{i} + 0\mathbf{j})$$

$$= 3\mathbf{i} + 1\mathbf{j}$$

$$Q = (3, 1).$$

Which is correct.

8.7 The Equation of a Line Segment

The line segment is very important in computer graphics as it is a fundamental
building block for 2-D and 3-D polygons, and all sorts of surface meshes. In $\mathbb{R}^2$
we need to clip line segments against rectangular windows; in $\mathbb{R}^3$ we need to clip
line segments against a viewing frustum; and in $\mathbb{R}^2$ and $\mathbb{R}^3$ we need to compute the
potential intersection of two line segments.

Fig. 8.22 Defining a line segment

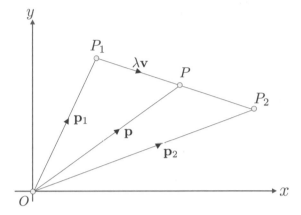

The parametric form of the straight-line equation is the most practical basis for manipulating straight-line segments, where the controlling parameter can be used to determine the position of a point along the segment.

Figure 8.22 shows the following conditions:

- $P_1(x_1, y_1)$ and $P_2(x_2, y_2)$ define the line segment,
- $\mathbf{p}_1 = x_1\mathbf{i} + y_1\mathbf{j}$ is P_1's position vector,
- $\mathbf{p}_2 = x_2\mathbf{i} + y_2\mathbf{j}$ is P_2's position vector,
- $P(x_p, y_p)$ is a point on the line segment,
- $\mathbf{p} = x_p\mathbf{i} + y_p\mathbf{j}$ is P's position vector,
- $\mathbf{v} = \overrightarrow{P_1P_2} = x_v\mathbf{i} + y_v\mathbf{j}$,
- λ is a scaling parameter for $\mathbf{v}$.

Therefore,

$$\mathbf{v} = \mathbf{p}_2 - \mathbf{p}_1$$
$$x_v = x_2 - x_1$$
$$y_v = y_2 - y_1$$
$$\mathbf{p} = \mathbf{p}_1 + \lambda\mathbf{v}$$
$$x_p = x_1 + \lambda(x_2 - x_1)$$
$$y_p = y_1 + \lambda(y_2 - y_1).$$

Point P is between P_1 and P_2 for $\lambda \in [0, 1]$. And by changing λ between 0 and 1, we slide along the line segment.

8.8   The Intersection of Two Straight Lines

The need to compute line intersections arises in 2-D and 3-D clipping, modeling and animation. Most of the time we are concerned with line segments, rather than infinite lines, although the latter are required in illumination calculations, ray casting and ray tracing. To set the scene, let us calculate the point of intersection of two straight lines in $\mathbb{R}^2$.

To begin, we must anticipate the two possibilities that can arise with two lines: the first is that the lines intersect, giving a single point of intersection; second, the line equations are linearly related, producing no intersections. Although one can approach this problem using line equations, we consider only a vector approach.

The two parametric lines are shown in Fig. 8.23 and are defined as follows:

- $R(x_r, y_r)$ is a point on the line with direction vector $\mathbf{a} = x_a\mathbf{i} + y_a\mathbf{j}$,
- $\mathbf{r} = x_r\mathbf{i} + y_r\mathbf{j}$ is R's position vector,
- $S(x_s, y_s)$ is a point on the line with direction vector $\mathbf{b} = x_b\mathbf{i} + y_b\mathbf{j}$,
- $\mathbf{s} = x_s\mathbf{i} + y_s\mathbf{j}$ is S's position vector,
- $P(x_p, y_p)$ is the point of intersection between the two lines,
- $\mathbf{p} = x_p\mathbf{i} + y_p\mathbf{j}$ is P's position vector,
- $\lambda\mathbf{a}$ is the vector $\overrightarrow{RP}$,
- $\epsilon\mathbf{b}$ is the vector $\overrightarrow{SP}$,

where,

$$\mathbf{p} = \mathbf{r} + \lambda\mathbf{a}$$
$$\mathbf{p} = \mathbf{s} + \epsilon\mathbf{b}.$$

The task now is to discover the values of λ and ϵ (epsilon) at the point P. Therefore,

$$\mathbf{r} + \lambda\mathbf{a} = \mathbf{s} + \epsilon\mathbf{b}. \tag{8.47}$$

Fig. 8.23 Two parametric lines

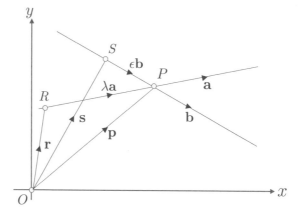

In order to isolate λ, we multiply (8.47) by $\mathbf{b}^\perp = -y_b\mathbf{i} + x_b\mathbf{j}$ using the dot product:

$$\mathbf{b}^\perp \cdot \mathbf{r} + \lambda\mathbf{b}^\perp \cdot \mathbf{a} = \mathbf{b}^\perp \cdot \mathbf{s} + \epsilon\mathbf{b}^\perp \cdot \mathbf{b}$$
$$\lambda\mathbf{b}^\perp \cdot \mathbf{a} = \mathbf{b}^\perp \cdot (\mathbf{s} - \mathbf{r}) + \epsilon\mathbf{b}^\perp \cdot \mathbf{b}. \tag{8.48}$$

But in (8.48) $\mathbf{b}^\perp \cdot \mathbf{b} = 0$, therefore,

$$\lambda\mathbf{b}^\perp \cdot \mathbf{a} = \mathbf{b}^\perp \cdot (\mathbf{s} - \mathbf{r})$$
$$\lambda = \frac{\mathbf{b}^\perp \cdot (\mathbf{s} - \mathbf{r})}{\mathbf{b}^\perp \cdot \mathbf{a}}$$

from which we can state:

$$\lambda = \frac{x_b(y_s - y_r) - y_b(x_s - x_r)}{x_b y_a - y_b x_a}. \tag{8.49}$$

In order to isolate ϵ we multiply (8.47) by $\mathbf{a}^\perp = -y_a\mathbf{i} + x_a\mathbf{j}$ using the dot product:

$$\mathbf{a}^\perp \cdot \mathbf{r} + \lambda\mathbf{a}^\perp \cdot \mathbf{a} = \mathbf{a}^\perp \cdot \mathbf{s} + \epsilon\mathbf{a}^\perp \cdot \mathbf{b}$$
$$\epsilon\mathbf{a}^\perp \cdot \mathbf{b} = \mathbf{a}^\perp \cdot (\mathbf{r} - \mathbf{s}) + \lambda\mathbf{a}^\perp \cdot \mathbf{a}. \tag{8.50}$$

But in (8.50) $\mathbf{a}^\perp \cdot \mathbf{a} = 0$, therefore,

$$\epsilon\mathbf{a}^\perp \cdot \mathbf{b} = \mathbf{a}^\perp \cdot (\mathbf{r} - \mathbf{s})$$
$$\epsilon = \frac{\mathbf{a}^\perp \cdot (\mathbf{r} - \mathbf{s})}{\mathbf{a}^\perp \cdot \mathbf{b}}$$

from which we can state:

$$\epsilon = \frac{x_a(y_r - y_s) - y_a(x_r - x_s)}{x_a y_b - y_a x_b}. \tag{8.51}$$

The coordinates of P are given by

$$x_p = x_r + \lambda x_a, \quad y_p = y_r + \lambda y_a$$

or

$$x_p = x_s + \epsilon x_b, \quad y_p = y_s + \epsilon y_b.$$

But what if the line equations are linearly related? Well, this can be detected by

$$\begin{vmatrix} x_a & y_a \\ x_b & y_b \end{vmatrix} = 0.$$

Let's test the above equations with an example.

Fig. 8.24 Two intersecting
parametric lines

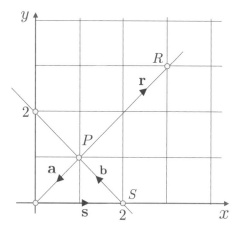

Figure 8.24 shows two lines intersecting at $P(1, 1)$, with the following conditions:

- $R(3, 3)$,
- $\mathbf{r} = 3\mathbf{i} + 3\mathbf{j}$,
- $S(2, 0)$,
- $\mathbf{s} = 2\mathbf{i} + 0\mathbf{j}$,
- $\mathbf{a} = -1\mathbf{i} - 1\mathbf{j}$,
- $\mathbf{b} = -1\mathbf{i} + 1\mathbf{j}$.

Therefore,

$$
\begin{aligned}
\lambda &= \frac{x_b(y_s - y_r) - y_b(x_s - x_r)}{x_b y_a - y_b x_a} \\
&= \frac{-1(0 - 3) - 1(2 - 3)}{(-1)(-1) - 1(-1)} \\
&= \frac{3 + 1}{2} = 2
\end{aligned}
$$

or,

$$
\begin{aligned}
\epsilon &= \frac{x_a(y_r - y_s) - y_a(x_r - x_s)}{x_a y_b - y_a x_b} \\
&= \frac{(-1)(3 - 0) - (-1)(3 - 2)}{(-1)1 - (-1)(-1)} \\
&= \frac{-3 + 1}{-2} = 1
\end{aligned}
$$

and

$$x_p = x_r + \lambda x_a = 3 + 2(-1) = 1$$
$$y_p = y_r + \lambda y_a = 3 + 2(-1) = 1$$

or,

$$x_p = x_s + \epsilon x_b = 2 + 1(-1) = 1$$
$$y_p = y_s + \epsilon y_b = 0 + 1(1) = 1$$

making the point of intersection $(1, 1)$, which is correct.

8.9   The Point of Intersection of Two Line Segments in $\mathbb{R}^2$

In the previous section we calculated the intersection point of two infinite 2-D lines, and in this section we exploit this knowledge to detect the potential intersection of two finite 2-D line segments. As the segments are represented by parametric vectors, the value of the parameter is the key to identifying intersections, non-intersections and touch conditions. Figure 8.25 shows two line segments with the following conditions:

- $R_1(x_{r1}, y_{r1})$ is the end of a line, with position vector $\mathbf{r}_1 = x_{r1}\mathbf{i} + y_{r1}\mathbf{j}$,
- $R_2(x_{r2}, y_{r2})$ is the end of a line, with position vector $\mathbf{r}_2 = x_{r2}\mathbf{i} + y_{r2}\mathbf{j}$,
- $S_1(x_{s1}, y_{s1})$ is the end of a line, with position vector $\mathbf{s}_1 = x_{s1}\mathbf{i} + y_{s1}\mathbf{j}$,
- $S_2(x_{s2}, y_{s2})$ is the end of a line, with position vector $\mathbf{s}_2 = x_{s2}\mathbf{i} + y_{s2}\mathbf{j}$,
- $P(x_p, y_p)$ is the point of intersection between the two lines,
- $\mathbf{p} = x_p\mathbf{i} + y_p\mathbf{j}$ is P's position vector,
- $\lambda\mathbf{a}$ is the vector $\overrightarrow{R_1P}$,
- $\epsilon\mathbf{b}$ is the vector $\overrightarrow{S_1P}$,

where,

$$\mathbf{p} = \mathbf{r}_1 + \lambda\mathbf{a}$$
$$\mathbf{p} = \mathbf{s}_1 + \epsilon\mathbf{b}.$$

For intersection at P:

$$\mathbf{r}_1 + \lambda\mathbf{a} = \mathbf{s}_1 + \epsilon\mathbf{b}. \tag{8.52}$$

In order to isolate λ, we multiply (8.52) by $\mathbf{b}^\perp = -y_b\mathbf{i} + x_b\mathbf{j}$ using the dot product:

$$\mathbf{b}^\perp \cdot \mathbf{r}_1 + \lambda\mathbf{b}^\perp \cdot \mathbf{a} = \mathbf{b}^\perp \cdot \mathbf{s}_1 + \epsilon\mathbf{b}^\perp \cdot \mathbf{b}$$
$$\lambda\mathbf{b}^\perp \cdot \mathbf{a} = \mathbf{b}^\perp \cdot (\mathbf{s}_1 - \mathbf{r}_1) + \epsilon\mathbf{b}^\perp \cdot \mathbf{b}. \tag{8.53}$$

Fig. 8.25 Two intersecting line segments

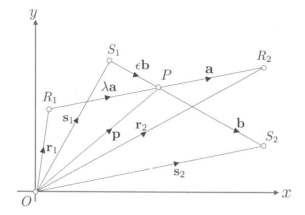

But in (8.53) $\mathbf{b}^\perp \cdot \mathbf{b} = 0$, therefore,

$$\lambda \mathbf{b}^\perp \cdot \mathbf{a} = \mathbf{b}^\perp \cdot (\mathbf{s}_1 - \mathbf{r}_1)$$

$$\lambda = \frac{\mathbf{b}^\perp \cdot (\mathbf{s}_1 - \mathbf{r}_1)}{\mathbf{b}^\perp \cdot \mathbf{a}}$$

from which we can state:

$$\lambda = \frac{x_b(y_{s1} - y_{r1}) - y_b(x_{s1} - x_{r1})}{x_b y_a - y_b x_a}. \tag{8.54}$$

In order to isolate ϵ, we multiply (8.52) by $\mathbf{a}^\perp = -y_a \mathbf{i} + x_a \mathbf{j}$ using the dot product:

$$\mathbf{a}^\perp \cdot \mathbf{r}_1 + \lambda \mathbf{a}^\perp \cdot \mathbf{a} = \mathbf{a}^\perp \cdot \mathbf{s}_1 + \epsilon \mathbf{a}^\perp \cdot \mathbf{b}$$

$$\epsilon \mathbf{a}^\perp \cdot \mathbf{b} = \mathbf{a}^\perp \cdot (\mathbf{r}_1 - \mathbf{s}_1) + \lambda \mathbf{a}^\perp \cdot \mathbf{a}. \tag{8.55}$$

But in (8.55) $\mathbf{a}^\perp \cdot \mathbf{a} = 0$, therefore,

$$\epsilon \mathbf{a}^\perp \cdot \mathbf{b} = \mathbf{a}^\perp \cdot (\mathbf{r}_1 - \mathbf{s}_1)$$

$$\epsilon = \frac{\mathbf{a}^\perp \cdot (\mathbf{r}_1 - \mathbf{s}_1)}{\mathbf{a}^\perp \cdot \mathbf{b}}$$

from which we can state:

$$\epsilon = \frac{x_a(y_{r1} - y_{s1}) - y_a(x_{r1} - x_{s1})}{x_a y_b - y_a x_b}. \tag{8.56}$$

Fig. 8.26 Two intersecting
line segments

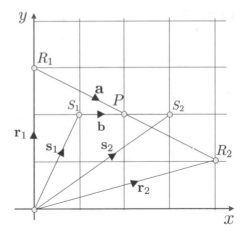

Only if $\lambda, \epsilon \in [0, 1]$, do the lines intersect or touch one another, and the coordinates of P are given by

$$x_p = x_{r1} + \lambda x_a, \quad y_p = y_{r1} + \lambda y_a$$

or,

$$x_p = x_{s1} + \epsilon x_b, \quad y_p = y_{s1} + \epsilon y_b.$$

Figure 8.26 shows two intersecting line segments at $P(2, 2)$, with the following conditions:

- $R_1(0, 3)$ and $R_2(4, 1)$ define the first line segment,
- $S_1(1, 2)$ and $S_2(3, 2)$ define the second line segment,
- $\mathbf{r}_1 = 0\mathbf{i} + 3\mathbf{j}$ and $\mathbf{r}_2 = 4\mathbf{i} + 1\mathbf{j}$ are position vectors,
- $\mathbf{s}_1 = 1\mathbf{i} + 2\mathbf{j}$ and $\mathbf{s}_2 = 3\mathbf{i} + 2\mathbf{j}$ are position vectors,
- $\mathbf{a} = 4\mathbf{i} - 2\mathbf{j}$ and $\mathbf{b} = 2\mathbf{i} + 0\mathbf{j}$ are direction vectors.

Therefore,

$$\begin{aligned}
\lambda &= \frac{x_b(y_{s1} - y_{r1}) - y_b(x_{s1} - x_{r1})}{x_b y_a - y_b x_a} \\
&= \frac{2(2 - 3) - 0(1 - 0)}{2(-2) - 0(4)} = \tfrac{1}{2} \\
x_p &= x_{r1} + \lambda x_a \\
&= 0 + \tfrac{1}{2}4 = 2 \\
y_p &= y_{r1} + \lambda y_a \\
&= 3 + \tfrac{1}{2}(-2) = 2
\end{aligned}$$

and,

$$\epsilon = \frac{x_a(y_{r1} - y_{s1}) - y_a(x_{r1} - x_{s1})}{x_a y_b - y_a x_b}$$

$$= \frac{4(3 - 2) - (-2)(0 - 1)}{4(0) - (-2)2} = \frac{1}{2}$$

$$x_p = x_{s1} + \epsilon x_b$$

$$= 1 + \tfrac{1}{2}2 = 2$$

$$y_p = y_{s1} + \epsilon y_b$$

$$= 2 + \tfrac{1}{2}(0) = 2.$$

Confirming that the point of intersection is $P(2, 2)$.

Now let's examine the case of two line segments that touch, as shown in Fig. 8.27 with the following conditions:

- $R_1(0, 3)$ and $R_2(4, 1)$ define the first line segment,
- $S_1(1, 3)$ and $S_2(4, 1)$ define the second line segment,
- $\mathbf{r}_1 = 0\mathbf{i} + 3\mathbf{j}$ and $\mathbf{r}_2 = 4\mathbf{i} + 1\mathbf{j}$ are position vectors,
- $\mathbf{s}_1 = 1\mathbf{i} + 3\mathbf{j}$ and $\mathbf{s}_2 = 4\mathbf{i} + 1\mathbf{j}$ are position vectors,
- $\mathbf{a} = 4\mathbf{i} - 2\mathbf{j}$ and $\mathbf{b} = 3\mathbf{i} - 2\mathbf{j}$ are direction vectors.

Therefore,

$$\lambda = \frac{x_b(y_{s1} - y_{r1}) - y_b(x_{s1} - x_{r1})}{x_b y_a - y_b x_a}$$

$$= \frac{3(3 - 3) - (-2)(1 - 0)}{3(-2) - (-2)(4)} = 1$$

$$x_p = x_{r1} + \lambda x_a$$

$$= 0 + 1(4) = 4$$

$$y_p = y_{r1} + \lambda y_a$$

$$= 3 + 1(-2) = 1$$

and,

$$\epsilon = \frac{x_a(y_{r1} - y_{s1}) - y_a(x_{r1} - x_{s1})}{x_a y_b - y_a x_b}$$

$$= \frac{4(3 - 3) - (-2)(0 - 1)}{4(-2) - (-2)3} = 1$$

$$x_p = x_{s1} + \epsilon x_b$$

$$= 1 + 1(3) = 4$$

$$y_p = y_{s1} + \epsilon y_b$$

$$= 3 + 1(-2) = 1.$$

Fig. 8.27 Two touching line
segments

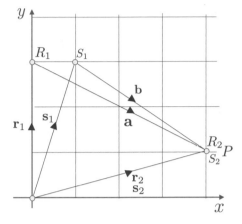

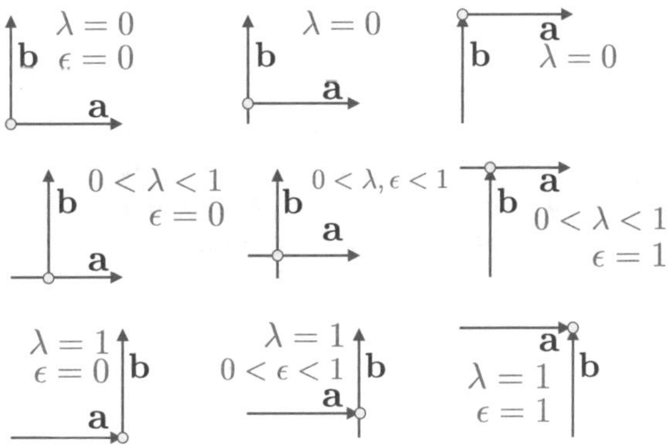

Fig. 8.28 Relative positions
of two line segments

Confirming that the line segments touch at $P(4, 1)$. The line segments are parallel
if the denominator in (8.54) or (8.56) is zero. Figure 8.28 illustrates the relative
positions of the line segments for different values of λ and ϵ.

8.10   Summary

This chapter has examined different ways of representing lines and line segments
using vectors, in various geometric scenarios. Hopefully you will have noticed that
I am a firm believer in drawing a clear diagram with all the vectors associated with
the scenario, as it helps reveal possible ways of solving the problem. One particular
problem-solving strategy is the use of perpendicular vectors in conjunction with the
dot product.

Chapter 9
The Plane

9.1 Introduction

The main objective of this chapter is to solve useful geometric problems involving the plane. We begin this exploration by examining the Cartesian form, and the parametric form of the plane equation. This is followed by a variety of problems that include: defining a plane from three points, 3-D space partitioning, the angle between two planes, the position and distance of a point to a plane, the reflection of a point in a plane, and lastly, a plane between two points. Each problem includes a diagram and a method of defining the equations that reveal a solution.

9.2 The Cartesian Form of the Plane Equation

The general form of the plane equation creates an equation that equals zero, whereas the Cartesian form arranges the equation such that a constant term is isolated on one side of the equals sign:

$$ax + by + cz = d.$$

Both forms have their individual advantages, but the Cartesian form is useful from a geometric perspective.

We use vector analysis to derive the plane equation and the reader will see that there is an intimate relationship between this and the Cartesian form of the line equation described in Chap. 8. Furthermore, we will approach the analysis in a similar fashion.

Step 1
Define a plane containing the point $P(x, y, z)$ and its associated position vector $\mathbf{p} = x\mathbf{i} + y\mathbf{j} + z\mathbf{k}$, away from the origin O, as shown in Fig. 9.1.

© Springer-Verlag London Ltd., part of Springer Nature 2021
J. Vince, *Vector Analysis for Computer Graphics*,
https://doi.org/10.1007/978-1-4471-7505-6_9

Fig. 9.1 Step 1

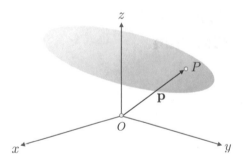

Fig. 9.2 Step 2

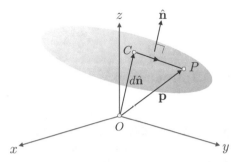

Step 2

Construct a line from the origin to a point C on the plane such that $\overrightarrow{OC}$ is perpendicular to the plane. It is convenient to make $\overrightarrow{OC}$ a scalar multiple of a unit vector $\hat{\mathbf{n}} = a\mathbf{i} + b\mathbf{j} + c\mathbf{k}$. i.e. $\overrightarrow{OC} = d\hat{\mathbf{n}}$, where d is some scalar as shown in Fig. 9.2. We can now write $\overrightarrow{OP} = \overrightarrow{OC} + \overrightarrow{CP}$. Substituting vector names gives

$$\mathbf{p} = d\hat{\mathbf{n}} + \overrightarrow{CP}. \tag{9.1}$$

Although we know d and $\hat{\mathbf{n}}$, and the fact that $\mathbf{p}$ points to any point on the plane, we do not know $\overrightarrow{CP}$. Somehow it has to be eliminated, which is achieved by multiplying (9.1) by $\hat{\mathbf{n}}$ using the dot product:

$$\hat{\mathbf{n}} \cdot \mathbf{p} = d\hat{\mathbf{n}} \cdot \hat{\mathbf{n}} + \hat{\mathbf{n}} \cdot \overrightarrow{CP}. \tag{9.2}$$

However, the dot product of two perpendicular vectors is zero. Therefore,

$$\hat{\mathbf{n}} \cdot \overrightarrow{CP} = 0$$

and (9.2) reduces to

$$\hat{\mathbf{n}} \cdot \mathbf{p} = d\hat{\mathbf{n}} \cdot \hat{\mathbf{n}}. \tag{9.3}$$

But we already know that $\hat{\mathbf{n}} \cdot \hat{\mathbf{n}} = 1$. Therefore,

$$\hat{\mathbf{n}} \cdot \mathbf{p} = d. \tag{9.4}$$

Expanding (9.4) we get

$$(a\mathbf{i} + b\mathbf{j} + c\mathbf{k}) \cdot (x\mathbf{i} + y\mathbf{j} + z\mathbf{k}) = d$$
$$ax + by + cz = d$$

which we recognise as the Cartesian form of the plane equation where:

x, y, z are the coordinates of a point on the plane,
a, b, c are the components of a unit vector normal to the plane,
d is the perpendicular distance from the origin to the plane.

The above analysis assumes that the vector normal to the plane is a unit vector, which gives a precise geometric meaning to d. However, we must be careful not to apply this geometric meaning to all plane equations. For example, consider (9.5):

$$2x + 3y + 4z = 10. \tag{9.5}$$

The vector normal to the plane is clearly not a unit vector as its magnitude is

$$\sqrt{2^2 + 3^2 + 4^2} = \sqrt{29}.$$

But if we divide (9.5) throughout by $\sqrt{29}$, we obtain (9.6):

$$\frac{2}{\sqrt{29}}x + \frac{3}{\sqrt{29}}y + \frac{4}{\sqrt{29}}z = \frac{10}{\sqrt{29}} \tag{9.6}$$

where the perpendicular distance from the origin to the plane is $\frac{10}{\sqrt{29}}$.

Let us now consider an alternative definition of the plane equation – the parametric form.

9.3 The Parametric Form of the Plane Equation

The parametric form of the plane equation requires two non-linearly connected vectors and a single point to define a plane. The two vectors provide the orientation of the plane, whilst the single point identifies a specific plane. Such a scenario is shown in Fig. 9.3, with the following points and vectors:

- $T(x_t, y_t, z_t)$ is the given point on the plane,
- $\mathbf{t} = x_t\mathbf{i} + y_t\mathbf{j} + z_t\mathbf{k}$ is T's position vector,
- $\mathbf{a} = x_a\mathbf{i} + y_a\mathbf{j} + z_a\mathbf{k}$ is the first vector parallel to the plane,

Fig. 9.3 The parametric
form of a plane

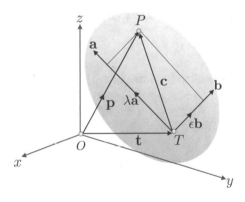

- $\mathbf{b} = x_b\mathbf{i} + y_b\mathbf{j} + z_b\mathbf{k}$ is the second vector parallel to the plane,
- $P(x, y, z)$ is any given point on the plane,
- $\mathbf{p} = x\mathbf{i} + y\mathbf{j} + z\mathbf{k}$ is P's position vector,
- $\mathbf{c} = \overrightarrow{TP}$,
- $\lambda\mathbf{a}$ is $\overrightarrow{TP}$'s projection on $\mathbf{a}$,
- $\epsilon\mathbf{b}$ is $\overrightarrow{TP}$'s projection on $\mathbf{b}$,
- λ and ϵ are scalars.

Therefore,

$$\mathbf{c} = \lambda\mathbf{a} + \epsilon\mathbf{b} \qquad (9.7)$$

and

$$\mathbf{p} = \mathbf{t} + \mathbf{c}. \qquad (9.8)$$

Therefore, combining (9.7) and (9.8)

$$\mathbf{p} = \mathbf{t} + \lambda\mathbf{a} + \epsilon\mathbf{b}. \qquad (9.9)$$

More explicitly:

$$x = x_t + \lambda x_a + \epsilon x_b$$
$$y = y_t + \lambda y_a + \epsilon y_b$$
$$z = x_t + \lambda z_a + \epsilon z_b.$$

If $\mathbf{a}$ and $\mathbf{b}$ are unit vectors, and are mutually perpendicular, i.e. $\mathbf{a} \cdot \mathbf{b} = 0$, λ and ϵ become linear measurements along the $\mathbf{a}$- and $\mathbf{b}$-axes relative to T.

Figure 9.4 illustrates a parametric plane with the following points and vectors:

- $T(2, 0, 0)$ is the given point on the plane,
- $\mathbf{a} = -2\mathbf{i} + 2\mathbf{j} + 0\mathbf{k}$ is the first vector parallel to the plane,
- $\mathbf{b} = -2\mathbf{i} + 0\mathbf{j} + 2\mathbf{k}$ is the second vector parallel to the plane,

Fig. 9.4 The parametric
form of a plane

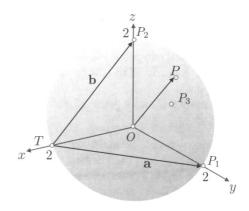

- $P(x, y, z)$ is any given point on the plane.

    Point T: $\lambda = 0$, $\epsilon = 0$:

$$x = 2 + 0 + 0 = 2$$
$$y = 0 + 0 + 0 = 0$$
$$z = 0 + 0 + 0 = 0.$$

Point P_1: $\lambda = 1$, $\epsilon = 0$:

$$x = 2 + 1(-2) + 0(-2) = 0$$
$$y = 0 + 1(2) + 0(0) = 2$$
$$z = 0 + 1(0) + 0(2) = 0.$$

Point P_2: $\lambda = 0$, $\epsilon = 1$:

$$x = 2 + 0(-2) + 1(-2) = 0$$
$$y = 0 + 0(2) + 1(0) = 0$$
$$z = 0 + 0(0) + 1(2) = 2.$$

Point P_3: $\lambda = \frac{1}{2}$, $\epsilon = \frac{1}{2}$:

$$x = 2 + \tfrac{1}{2}(-2) + \tfrac{1}{2}(-2) = 0$$
$$y = 0 + \tfrac{1}{2}(2) + \tfrac{1}{2}(0) = 1$$
$$z = 0 + \tfrac{1}{2}(0) + \tfrac{1}{2}(2) = 1.$$

9.4   A Plane Equation from Three Points

We know that two points are required to define a line, but three points are needed to define a plane. So let us now consider how the plane equation is derived from three such points.

Figure 9.5 shows a plane defined by three points with the following points and vectors:

- $R(x_r, y_r, z_r)$ is a point defining the plane,
- $S(x_s, y_s, z_s)$ is a point defining the plane,
- $T(x_t, y_t, z_t)$ is a point defining the plane,
- $P(x, y, z)$ is any point on the plane,
- $\overrightarrow{RS} = \mathbf{u} = x_u \mathbf{i} + y_u \mathbf{j} + z_u \mathbf{k}$ is a vector parallel to the plane,
- $\overrightarrow{RT} = \mathbf{v} = x_v \mathbf{i} + y_v \mathbf{j} + z_v \mathbf{k}$ is a vector parallel to the plane,
- $\overrightarrow{RP} = \mathbf{w} = x_w \mathbf{i} + y_w \mathbf{j} + z_w \mathbf{k}$ is a vector parallel to the plane,
- $\mathbf{u} \times \mathbf{v}$ is a vector normal to the plane.

The three points are assumed to be in a counter-clockwise sequence viewed from the direction of the surface normal. The vector product $\mathbf{u} \times \mathbf{v}$ provides a vector normal to the plane containing the points:

$$\mathbf{u} \times \mathbf{v} = \begin{vmatrix} \mathbf{i} & \mathbf{j} & \mathbf{k} \\ x_u & y_u & z_u \\ x_v & y_v & z_v \end{vmatrix}.$$

We now take the point P on the plane and form the vector $\mathbf{w} = \overrightarrow{RP}$. Therefore, the dot product $\mathbf{w} \cdot (\mathbf{u} \times \mathbf{v}) = 0$. This condition can be expressed as a determinant and converted into the general equation of a plane:

Fig. 9.5 Three points R, S, T defining a plane

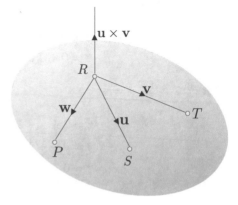

$$\mathbf{w} \cdot (\mathbf{u} \times \mathbf{v}) = \begin{vmatrix} x_w & y_w & z_w \\ x_u & y_u & z_u \\ x_v & y_v & z_v \end{vmatrix} = 0. \tag{9.10}$$

Expanding (9.10), we obtain (9.11):

$$x_w \begin{vmatrix} y_u & z_u \\ y_v & z_v \end{vmatrix} + y_w \begin{vmatrix} z_u & x_u \\ z_v & x_v \end{vmatrix} + z_w \begin{vmatrix} x_u & y_u \\ x_v & y_v \end{vmatrix} = 0 \tag{9.11}$$

which becomes (9.12):

$$(x - x_r) \begin{vmatrix} y_s - y_r & z_s - z_r \\ y_t - y_r & z_t - z_r \end{vmatrix}$$

$$+ (y - y_r) \begin{vmatrix} z_s - z_r & x_s - x_r \\ z_t - z_r & x_t - x_r \end{vmatrix}$$

$$+ (z - z_r) \begin{vmatrix} x_s - x_r & y_s - y_r \\ x_t - x_r & y_t - y_r \end{vmatrix} = 0. \tag{9.12}$$

Equation (9.12) can be arranged in the Cartesian form of the line equation $ax + by + cz = d$, where,

$$a = \begin{vmatrix} y_s - y_r & z_s - z_r \\ y_t - y_r & z_t - z_r \end{vmatrix} \quad \text{or} \quad a = \begin{vmatrix} 1 & y_r & z_r \\ 1 & y_s & z_s \\ 1 & y_t & z_t \end{vmatrix} \tag{9.13}$$

$$b = \begin{vmatrix} z_s - z_r & x_s - x_r \\ z_t - z_r & x_t - x_r \end{vmatrix} \quad \text{or} \quad b = \begin{vmatrix} x_r & 1 & z_r \\ x_s & 1 & z_s \\ x_t & 1 & z_t \end{vmatrix} \tag{9.14}$$

$$c = \begin{vmatrix} x_s - x_r & y_s - y_r \\ x_t - x_r & y_t - y_r \end{vmatrix} \quad \text{or} \quad c = \begin{vmatrix} x_r & y_r & 1 \\ x_s & y_s & 1 \\ x_t & y_t & 1 \end{vmatrix} \tag{9.15}$$

$$d = ax_r + by_r + cz_r. \tag{9.16}$$

Let's test these results with an example. Figure 9.6 shows a plane with the following points and vectors:

- $R(2, 0, 0)$ is a point defining the plane,
- $S(0, 2, 0)$ is a point defining the plane,
- $T(0, 0, 2)$ is a point defining the plane,
- $P(x, y, z)$ is any point on the plane.

Fig. 9.6 Three points R, S, T, defining a plane

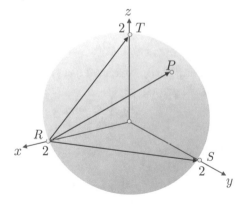

Therefore, substituting values into (9.13), (9.14), (9.15) and (9.16), we have:

$$a = \begin{vmatrix} 1 & y_r & z_r \\ 1 & y_s & z_s \\ 1 & y_t & z_t \end{vmatrix} = \begin{vmatrix} 1 & 0 & 0 \\ 1 & 2 & 0 \\ 1 & 0 & 2 \end{vmatrix} = 4$$

$$b = \begin{vmatrix} x_r & 1 & z_r \\ x_s & 1 & z_s \\ x_t & 1 & z_t \end{vmatrix} = \begin{vmatrix} 2 & 1 & 0 \\ 0 & 1 & 0 \\ 0 & 1 & 2 \end{vmatrix} = 4$$

$$c = \begin{vmatrix} x_r & y_r & 1 \\ x_s & y_s & 1 \\ x_t & y_t & 1 \end{vmatrix} = \begin{vmatrix} 2 & 0 & 1 \\ 0 & 2 & 1 \\ 0 & 0 & 1 \end{vmatrix} = 4$$

$$d = ax_r + by_r + cz_r = 4 \times 2 + 4 \times 0 + 4 \times 0 = 8$$

which makes the plane equation:

$$4x + 4y + 4z = 8.$$

9.5   3-D Space Partitioning

In Section 8.3 we discovered how it is possible to determine whether a point is inside or outside a convex boundary. The technique also reveals whether a point is on a boundary vertex or edge. Now let's explore a similar technique where we can determine whether a point is inside, outside, or on a vertex, edge or surface forming a convex polyhedron.

Fig. 9.7 Six views of a unit
cube

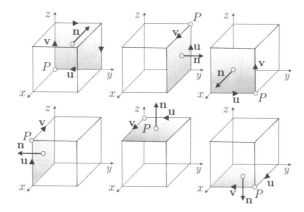

A simple example reveals the underlying mechanism of this form of 3-D space partitioning. A unit cube is chosen aligned with the x-, y- and z-axis, as it keeps the mathematics very simple. However, it should be clear from this example that the mathematical reasoning holds for any convex polyhedron.

Figure 9.7 shows six views of a unit cube with each surface highlighted with its normal vector $\mathbf{n}$. To ensure that all the normal vectors point outwards, it is derived from two similar edges forming each surface. We assume that the surfaces are defined by a chain of edges defined in a counter-clockwise sequence when viewed from the outside, as shown in Fig. 9.7.

The cross-product operation is used to create a normal vector from two adjacent edges selected from each surface. Any two will do, but the selection procedure must be the same for all the surfaces, otherwise the normal vector will not point consistently inwards or outwards. The chosen edges are the vectors $\mathbf{u}$ and $\mathbf{v}$, as shown in Fig. 9.7.

$$\mathbf{n} = \begin{vmatrix} \mathbf{i} & \mathbf{j} & \mathbf{k} \\ x_u & y_u & z_u \\ x_v & y_v & z_v \end{vmatrix}.$$

Furthermore, given a vector

$$\mathbf{n} = a\mathbf{i} + b\mathbf{j} + c\mathbf{k}$$

the plane equation is given by

$$ax + by + cz - (ax_p + by_p + cz_p) = 0$$

where $P(x_p, y_p, z_p)$ is a point on the plane. Such a point is shown on each view in Fig. 9.7. For example, in the first view:

$$\mathbf{u} = -\mathbf{j}, \quad \mathbf{v} = \mathbf{k}, \quad P = (0, 0, 0)$$

$$n = \begin{vmatrix} i & j & k \\ 0 & -1 & 0 \\ 0 & 0 & 1 \end{vmatrix} = -i$$

therefore, the plane equation is:

$$-x = 0.$$

In the second view:

$$u = k, \quad v = i, \quad P = (0, 1, 1)$$

$$n = \begin{vmatrix} i & j & k \\ 0 & 0 & 1 \\ 1 & 0 & 0 \end{vmatrix} = j$$

therefore, the plane equation is:

$$y - 1 = 0.$$

In the third view:

$$u = j, \quad v = k, \quad P = (1, 1, 0)$$

$$n = \begin{vmatrix} i & j & k \\ 0 & 1 & 0 \\ 0 & 0 & 1 \end{vmatrix} = i$$

therefore, the plane equation is:

$$x - 1 = 0.$$

In the fourth view:

$$u = k, \quad v = -i, \quad P = (1, 0, 1)$$

$$n = \begin{vmatrix} i & j & k \\ 0 & 0 & 1 \\ -1 & 0 & 0 \end{vmatrix} = -j$$

therefore, the plane equation is:

$$-y = 0.$$

In the fifth view:

$$u = -j, \quad v = i, \quad P = (0, 0, 1)$$

$$n = \begin{vmatrix} i & j & k \\ 0 & -1 & 0 \\ 1 & 0 & 0 \end{vmatrix} = k$$

therefore, the plane equation is:

$$z - 1 = 0.$$

In the sixth view:

$$\mathbf{u} = \mathbf{i}, \quad \mathbf{v} = -\mathbf{j}, \quad P = (1, 1, 0)$$

$$\mathbf{n} = \begin{vmatrix} \mathbf{i} & \mathbf{j} & \mathbf{k} \\ 1 & 0 & 0 \\ 0 & -1 & 0 \end{vmatrix} = -\mathbf{k}$$

therefore, the plane equation is:

$$-z = 0.$$

These results are shown in Table 9.1.

The LHS of the plane equation returns a value of zero for any point on the plane; a positive or negative value is returned for points not on the plane. If the point is in the space partition occupied by the surface normal, a positive value is returned, otherwise it is negative. Table 9.2 illustrates how the LHS expressions of the plane equations react to five different points. With reference to Fig. 9.7 the point $\left(\frac{1}{2}, \frac{1}{2}, \frac{1}{2}\right)$ is clearly inside the convex volume, and all the expressions are negative. Whereas, the point $(1, 0, 2)$ is outside the volume and the expression $z - 1$, returns a positive value. Consequently, if any expression goes positive, the search can halt and the point is declared outside.

If a point resides on one of the surfaces the corresponding expression returns a zero value, as shown with the point $\left(\frac{1}{2}, 1, \frac{1}{2}\right)$. If a point resides on an edge, two expressions return a zero value, as shown with the point $\left(\frac{1}{2}, 0, 1\right)$. Finally, if a point resides on a vertex, three expressions return a zero value, as shown with the point $(1, 1, 1)$.

Hopefully, this simple exercise has revealed just how useful vectors are in resolving a very useful space partitioning technique.

Table 9.1 Summary of vectors and plane equations

Surface	u	v	n	P	Equation
1	$-\mathbf{j}$	$\mathbf{k}$	$-\mathbf{i}$	$(0, 0, 0)$	$-x = 0$
2	$\mathbf{k}$	$\mathbf{i}$	$\mathbf{j}$	$(0, 1, 1)$	$y - 1 = 0$
3	$\mathbf{j}$	$\mathbf{k}$	$\mathbf{i}$	$(1, 1, 0)$	$x - 1 = 0$
4	$\mathbf{k}$	$-\mathbf{i}$	$\mathbf{j}$	$(1, 0, 1)$	$-y = 0$
5	$-\mathbf{j}$	$\mathbf{i}$	$\mathbf{k}$	$(0, 0, 1)$	$z - 1 = 0$
6	$\mathbf{i}$	$-\mathbf{j}$	$-\mathbf{k}$	$(1, 1, 0)$	$-z = 0$

Table 9.2 Partition for five points

LHS expression	$\left(\frac{1}{2}, \frac{1}{2}, \frac{1}{2}\right)$	(1, 0, 2)	$\left(\frac{1}{2}, 1, \frac{1}{2}\right)$	$\left(\frac{1}{2}, 0, 1\right)$	(1, 1, 1)
$-x$	$-$	$-$	$-$	$-$	$-$
$y - 1$	$-$	$-$	0	$-$	0
$x - 1$	$-$	0	$-$	$-$	0
$-y$	$-$	0	$-$	0	$-$
$z - 1$	$-$	$+$	$-$	0	0
$-z$	$-$	$-$	$-$	$-$	$-$
Result	Inside	Outside	Surface	Edge	Vertex

9.6   The Angle Between Two Planes

The angle between two planes is readily computed using the dot product of the two surface normals. For example, given two Cartesian plane equations:

$$a_1 x + b_1 y + c_1 z = d_1$$
$$a_2 x + b_2 y + c_2 z = d_2.$$

The normal vectors are

$$\mathbf{n}_1 = a_1 \mathbf{i} + b_1 \mathbf{j} + c_1 \mathbf{k}$$
$$\mathbf{n}_2 = a_2 \mathbf{i} + b_2 \mathbf{j} + c_2 \mathbf{k}.$$

Therefore,

$$\mathbf{n}_1 \cdot \mathbf{n}_2 = \|\mathbf{n}_1\| \, \|\mathbf{n}_2\| \cos \alpha$$

and

$$\alpha = \cos^{-1}\left(\frac{\mathbf{n}_1 \cdot \mathbf{n}_2}{\|\mathbf{n}_1\| \, \|\mathbf{n}_2\|}\right).$$

For example, Fig. 9.8 shows two planes whose plane equations are given by:

$$z = 0, \quad \text{and} \quad y + z = 1.$$

Therefore, the two normal vectors are

$$\mathbf{n}_1 = \mathbf{k}, \quad \text{and} \quad \mathbf{n}_2 = \mathbf{j} + \mathbf{k}.$$

Therefore,

$$\alpha = \cos^{-1}\left(\frac{\mathbf{k} \cdot (\mathbf{j} + \mathbf{k})}{1 \times \sqrt{2}}\right) = \cos^{-1}\left(\frac{1}{\sqrt{2}}\right) = 45°.$$

Fig. 9.8 The angle between two planes

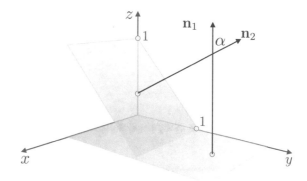

9.7 The Position and Distance of the Nearest Point on a Plane to a Point

This problem is concerned with the relationship between a point and a plane; in particular, the problem of finding the position and distance of a point on a plane that is the nearest to some specific point. For example, Fig. 9.9 shows a plane and a point P in space. The problem is to find the location and distance of a point on the plane nearest to P. Hopefully, it is obvious that the nearest point on the plane is perpendicular to P, which means that the dot product plays some role in the solution. Let's examine how vector analysis reveals a solution.

Figure 9.9 shows the following points and vectors:

- $ax + by + cz = d$ is the plane's equation,
- $\mathbf{n} = a\mathbf{i} + b\mathbf{j} + c\mathbf{k}$ is the plane's normal vector,
- $P(x_p, y_p, z_p)$ is a point away from the plane,
- $\mathbf{p} = x_p\mathbf{i} + y_p\mathbf{j} + z_p\mathbf{k}$ is P's position vector,
- $Q(x, y, z)$ is the nearest point on the plane to P,
- $\mathbf{q} = x\mathbf{i} + y\mathbf{j} + z\mathbf{k}$ is Q's position vector,
- $\overrightarrow{QP} = \lambda\mathbf{n}$ is the distance from the plane to P,
- λ is a scalar.

Fig. 9.9 The point P and a plane

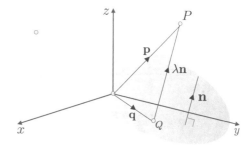

Therefore,

$$\mathbf{n} \cdot \mathbf{q} = (a\mathbf{i} + b\mathbf{j} + c\mathbf{k}) \cdot (x\mathbf{i} + y\mathbf{j} + z\mathbf{k}) = ax + by + cz = d. \qquad (9.17)$$

From Fig. 9.9,

$$\mathbf{q} = \mathbf{p} - \lambda\mathbf{n}. \qquad (9.18)$$

Multiply (9.18) by $\mathbf{n}$ using the dot product:

$$\mathbf{n} \cdot \mathbf{q} = \mathbf{n} \cdot \mathbf{p} - \lambda\mathbf{n} \cdot \mathbf{n}. \qquad (9.19)$$

Substitute (9.17) in (9.19):

$$d = \mathbf{n} \cdot \mathbf{p} - \lambda\mathbf{n} \cdot \mathbf{n}$$

and

$$\lambda = \frac{\mathbf{n} \cdot \mathbf{p} - d}{\mathbf{n} \cdot \mathbf{n}} = \frac{\mathbf{n} \cdot \mathbf{p} - d}{\|\mathbf{n}\|^2}.$$

Q is given by (9.18), and the distance of Q to P is

$$\|\overrightarrow{QP}\| = \|\lambda\mathbf{n}\|.$$

If $\mathbf{n}$ is a unit vector, then

$$\lambda = \hat{\mathbf{n}} \cdot \mathbf{p} - d$$

and

$$\|\overrightarrow{QP}\| = \lambda.$$

Let's test (9.18) with an example.

Figure 9.10 shows a plane whose equation is

$$x + y + z = 2.$$

Therefore,

$$\mathbf{n} = 1\mathbf{i} + 1\mathbf{j} + 1\mathbf{k}$$

and

$$d = 2.$$

We will find the nearest point on the plane to the origin by making $P = (0, 0, 0)$. Therefore,

$$\lambda = \frac{\mathbf{n} \cdot \mathbf{p} - d}{\|\mathbf{n}\|^2} = \frac{(1\mathbf{i} + 1\mathbf{j} + 1\mathbf{k}) \cdot (0\mathbf{i} + 0\mathbf{j} + 0\mathbf{k}) - 2}{\sqrt{3}\sqrt{3}} = -\frac{2}{3}$$

Fig. 9.10 The point P and a plane

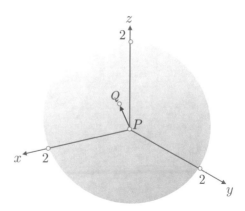

and

$$\mathbf{q} = \mathbf{p} - \lambda\mathbf{n} = (0\mathbf{i} + 0\mathbf{j} + 0\mathbf{k}) - \left(-\tfrac{2}{3}\right)(1\mathbf{i} + 1\mathbf{j} + 1\mathbf{k}) = \tfrac{2}{3}\mathbf{i} + \tfrac{2}{3}\mathbf{j} + \tfrac{2}{3}\mathbf{k}.$$

Therefore, the nearest point to $P(0, 0, 0)$ is $\left(\tfrac{2}{3}, \tfrac{2}{3}, \tfrac{2}{3}\right)$, and the distance is $\|\lambda\mathbf{n}\| = \tfrac{2}{3}\sqrt{3}$.

Now let's place P at $(2, 2, 2)$. The nearest point should still be Q, but the distance is $2\sqrt{3} - \tfrac{2}{3}\sqrt{3} = \tfrac{4}{3}\sqrt{3}$.

$$\lambda = \frac{\mathbf{n} \cdot \mathbf{p} - d}{\|\mathbf{n}\|^2} = \frac{(1\mathbf{i} + 1\mathbf{j} + 1\mathbf{k}) \cdot (2\mathbf{i} + 2\mathbf{j} + 2\mathbf{k}) - 2}{\sqrt{3}\sqrt{3}} = \tfrac{4}{3}$$

and

$$\mathbf{q} = \mathbf{p} - \lambda\mathbf{n} = (2\mathbf{i} + 2\mathbf{j} + 2\mathbf{k}) - \left(\tfrac{4}{3}\right)(1\mathbf{i} + 1\mathbf{j} + 1\mathbf{k}) = \tfrac{2}{3}\mathbf{i} + \tfrac{2}{3}\mathbf{j} + \tfrac{2}{3}\mathbf{k}.$$

Therefore, the nearest point to $P(2, 2, 2)$ is $\left(\tfrac{2}{3}, \tfrac{2}{3}, \tfrac{2}{3}\right)$, and the distance is $\|\lambda\mathbf{n}\| = \tfrac{4}{3}\sqrt{3}$, as predicted.

9.8 The Reflection of a Point in a Plane

When dealing with mirrors we need to know how to compute the virtual position of an object's reflection. Once we know how to compute a single point, most objects can be processed. Central to the solution of this problem is the fact that the virtual reflection appears the same distance behind the mirror as the object is in front of

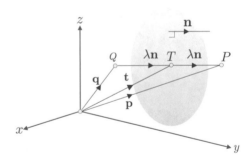

Fig. 9.11 *P*'s reflection in a plane

the mirror. Furthermore, a line connecting a real point to its virtual reflection is orthogonal to the mirror.

Figure 9.11 shows the following points and vectors:

- $ax + by + cz = d$ is the plane's equation,
- $\mathbf{n} = a\mathbf{i} + b\mathbf{j} + c\mathbf{k}$ is the plane's normal vector,
- $P(x_p, y_p, z_p)$ is the point to be reflected,
- $\mathbf{p} = x_p\mathbf{i} + y_p\mathbf{j} + z_p\mathbf{k}$ is P's position vector,
- $Q(x_q, y_q, z_q)$ is the reflection of P,
- $\mathbf{q} = x_q\mathbf{i} + y_q\mathbf{j} + z_q\mathbf{k}$ is Q's position vector,
- $T(x, y, z)$ is a point on the plane such that $\overrightarrow{TP}$ is parallel to $\mathbf{n}$,
- $\mathbf{t} = x\mathbf{i} + y\mathbf{j} + z\mathbf{k}$ is T's position vector,
- $\overrightarrow{QT} = \overrightarrow{TP} = \lambda\mathbf{n}$ is the distance from the plane to P,
- λ is a scalar.

Therefore,

$$\mathbf{n} \cdot \mathbf{t} = ax + by + cz = d. \tag{9.20}$$

From Fig. 9.11 we see that

$$\mathbf{p} = \mathbf{t} + \lambda\mathbf{n}. \tag{9.21}$$

Multiply (9.21) by $\mathbf{n}$ using the dot product:

$$\mathbf{n} \cdot \mathbf{p} = \mathbf{n} \cdot \mathbf{t} + \lambda\mathbf{n} \cdot \mathbf{n}. \tag{9.22}$$

Substitute (9.20) in (9.22)

$$\mathbf{n} \cdot \mathbf{p} = d + \lambda\mathbf{n} \cdot \mathbf{n}. \tag{9.23}$$

Rearrange (9.23):

$$\lambda = \frac{\mathbf{n} \cdot \mathbf{p} - d}{\mathbf{n} \cdot \mathbf{n}} = \frac{\mathbf{n} \cdot \mathbf{p} - d}{\|\mathbf{n}\|^2}. \tag{9.24}$$

From Fig. 9.11 we see that

$$\mathbf{q} = \mathbf{p} - 2\lambda\mathbf{n}. \tag{9.25}$$

Fig. 9.12 *P*'s reflection in a
plane

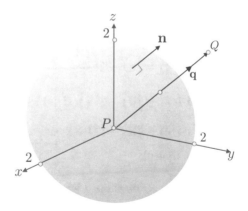

Therefore,

$$\mathbf{q} = \mathbf{p} - 2\left(\frac{\mathbf{n} \cdot \mathbf{p} - d}{\|\mathbf{n}\|^2}\right)\mathbf{n}. \tag{9.26}$$

If **n** is a unit vector, then

$$\mathbf{q} = \mathbf{p} - 2(\hat{\mathbf{n}} \cdot \mathbf{p} - d)\hat{\mathbf{n}}. \tag{9.27}$$

Let's explore two scenarios for these equations: one with a point behind the plane
and one in front, which will demonstrate that planes possess two reflecting sides.
Figure 9.12 shows a plane with the following points and vectors:

- $x + y + z = 2$ is the plane's equation,
- $\mathbf{n} = 1\mathbf{i} + 1\mathbf{j} + 1\mathbf{k}$ is the plane's normal vector,
- $P(0, 0, 0)$ is the point to be reflected,
- $\mathbf{p} = 0\mathbf{i} + 0\mathbf{j} + 0\mathbf{k}$ is P's position vector,
- $Q(x_q, y_q, z_q)$ is the reflection of P,
- $\mathbf{q} = x_q\mathbf{i} + y_q\mathbf{j} + z_q\mathbf{k}$ is Q's position vector.

Using (9.26):

$$\mathbf{q} = (0\mathbf{i} + 0\mathbf{j} + 0\mathbf{k}) - 2\left(\frac{(1\mathbf{i} + 1\mathbf{j} + 1\mathbf{k}) \cdot (0\mathbf{i} + 0\mathbf{j} + 0\mathbf{k}) - 2}{\sqrt{3}\sqrt{3}}\right)(1\mathbf{i} + 1\mathbf{j} + 1\mathbf{k})$$

$$= -2\left(\tfrac{-2}{3}\right)(1\mathbf{i} + 1\mathbf{j} + 1\mathbf{k})$$

$$= \tfrac{4}{3}\mathbf{i} + \tfrac{4}{3}\mathbf{j} + \tfrac{4}{3}\mathbf{k}.$$

If we now place P at $\left(\tfrac{4}{3}\mathbf{i} + \tfrac{4}{3}\mathbf{j} + \tfrac{4}{3}\mathbf{k}\right)$, its reflection should be at the origin.

Using (9.26):

$$q = \left(\tfrac{4}{3}i + \tfrac{4}{3}j + \tfrac{4}{3}k\right) - 2\left(\frac{(1i + 1j + 1k) \cdot \left(\tfrac{4}{3}i + \tfrac{4}{3}j + \tfrac{4}{3}k\right) - 2}{\sqrt{3}\sqrt{3}}\right)(1i + 1j + 1k)$$

$$= \left(\tfrac{4}{3}i + \tfrac{4}{3}j + \tfrac{4}{3}k\right) - 2\left(\tfrac{2}{3}\right)(1i + 1j + 1k)$$

$$= \left(\tfrac{4}{3}i + \tfrac{4}{3}j + \tfrac{4}{3}k\right) - \left(\tfrac{4}{3}i + \tfrac{4}{3}j + \tfrac{4}{3}k\right)$$

$$= 0i + 0j + 0k$$

making the reflection at the origin.

9.9 A Plane Between Two Points

Given two points P_1 and P_2, we can connect a straight line between them, and locate a plane orthogonal to this line and a specified distance from P_1. Let's examine how such a plane is identified.

Figure 9.13 shows two points P_1 and P_2 joined by a connecting line, with the following points and vectors:

- $ax + by + cz = d$ is the plane's equation,
- $n = ai + bj + ck$ is the plane's normal,
- $P_1(x_1, y_1, z_1)$ is one of the points,
- $p_1 = x_1i + y_1j + z_1k$ is P_1's position vector,
- $P_2(x_2, y_2, z_2)$ is one of the points,
- $p_2 = x_2i + y_2j + z_2k$ is P_2's position vector,
- $Q(x, y, z)$ is a point on the plane located λn from P_1,
- $q = xi + yj + zk$ is Q's position vector,
- $\overrightarrow{P_1Q} = \lambda n$,
- λ is a scalar.

The plane's normal vector is

$$n = p_2 - p_1. \tag{9.28}$$

Let Q be a point on $\overrightarrow{P_1P_2}$, such that

$$q = p_1 + \lambda n. \tag{9.29}$$

Therefore,

$$n \cdot q = d. \tag{9.30}$$

Fig. 9.13 A plane between
two points

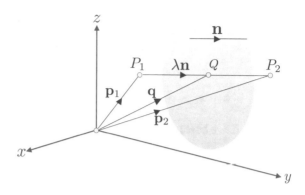

Fig. 9.14 A plane between
P_1 and P_2

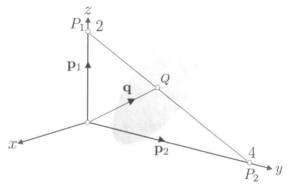

Substitute (9.29) in (9.30):

$$d = \mathbf{n} \cdot (\mathbf{p}_1 + \lambda\mathbf{n}) = \mathbf{n} \cdot \mathbf{p}_1 + \lambda\|\mathbf{n}\|^2. \tag{9.31}$$

Finally:

$$a = x_2 - x_1, \quad b = y_2 - y_1, \quad c = z_2 - z_1.$$

Let's test the above equations with the scenario shown in Fig. 9.14, with the following conditions:

- $\mathbf{p}_1 = 0\mathbf{i} + 0\mathbf{j} + 2\mathbf{k}$ is P_1's position vector,
- $P_2(0, 4, 0)$ is one of the points,
- $\mathbf{p}_2 = 0\mathbf{i} + 4\mathbf{j} + 0\mathbf{k}$ is P_2's position vector,
- $ax + by + cz = d$ is the plane's equation,
- $\mathbf{n} = a\mathbf{i} + b\mathbf{j} + c\mathbf{k}$ is the plane's normal,
- $Q(x, y, z)$ is a point on the plane located $\lambda\mathbf{n}$ from P_1,
- $\mathbf{q} = x\mathbf{i} + y\mathbf{j} + z\mathbf{k}$ is Q's position vector.

Let's find the plane equations for $\lambda = 0, \ 0.5, \ 1$.

Therefore,

$$\mathbf{n} = (0\mathbf{i} + 4\mathbf{j} - 2\mathbf{k})$$
$$\|\mathbf{n}\|^2 = 20$$

$$a = x_2 - x_1 = 0, \quad b = y_2 - y_1 = 4, \quad c = z_2 - z_1 = -2.$$

With $\lambda = 0$, we have:

$$d = (0\mathbf{i} + 4\mathbf{j} - 2\mathbf{k}) \cdot (0\mathbf{i} + 0\mathbf{j} + 2\mathbf{k}) + 0 \times 20 = -4$$

therefore, the plane equation is

$$4y - 2z = -4.$$

With $\lambda = 0.5$, we have:

$$d = (0\mathbf{i} + 4\mathbf{j} - 2\mathbf{k}) \cdot (0\mathbf{i} + 0\mathbf{j} + 2\mathbf{k}) + 0.5 \times 20 = 6$$

therefore, the plane equation is
$$4y - 2z = 6.$$

With $\lambda = 1$, we have:

$$d = (0\mathbf{i} + 4\mathbf{j} - 2\mathbf{k}) \cdot (0\mathbf{i} + 0\mathbf{j} + 2\mathbf{k}) + 1 \times 20 = 16$$

therefore, the plane equation is

$$4y - 2z = 16.$$

9.10   Summary

In this chapter we have shown a vectorial basis for the plane equation, and explored a variety of geometric problems involving a plane. Hopefully, this should enable the reader to solve similar problems with vectors.

Chapter 10
Intersections

10.1 Introduction

In this chapter we explore how vector-based techniques are used to resolve a variety of problems that arise in computer graphics such as proximity testing, modeling, collision detection, ray casting and ray tracing.

In some cases, as we shall see, we can work directly with vector-based equations without incorporating extra transformations. For example, to compute the intersection of two lines we can work directly with the line equations. But say we wish to compute the intersection of a cylinder with a line? Then, it is highly likely that the cylinder will have to be scaled, rotated and translated away from the origin. In which case, the cylinder's equations have to be transformed before performing the intersection. Unfortunately, the resulting solution is neither simple nor elegant. However, it just so happens that if we transform the line instead such that the relative orientation between the line and cylinder is preserved, the geometric analysis remains both simple and elegant.

For instance, let's assume that an object O is modeled such that its center is at the origin, and is then subjected to three transforms: $\mathbf{S}$ (scale), $\mathbf{R}$ (rotate) and $\mathbf{T}$ (translate) to produce the transformed object O':

$$O' = \mathbf{TRS}O.$$

If we attempt to compute an intersection with O' using a line described by

$$\mathbf{p} = \mathbf{t} + \lambda\mathbf{v}$$

the algebraic analysis becomes rather convoluted. However, if we leave the object where it is and transform the line instead, all that we have to make sure is that the line is transformed such that the relative orientation between the line and the object is preserved. Basically, this requires finding the inverse transforms of $\mathbf{T}$, $\mathbf{R}$ and $\mathbf{S}$:

© Springer-Verlag London Ltd., part of Springer Nature 2021
J. Vince, *Vector Analysis for Computer Graphics*,
https://doi.org/10.1007/978-1-4471-7505-6_10

$$\mathbf{T}^{-1}, \mathbf{R}^{-1}, \mathbf{S}^{-1}$$

and applying them to the line. But before so doing, we must remember that vectors are rather sensitive to transforms.

No attempt is made in the following sections to develop fast algorithms. The underlying goal is to identify strategies and to reveal patterns in solutions. Let's begin this survey with the simple case of two intersecting lines in $\mathbb{R}^2$.

10.2 Two Intersecting Lines in $\mathbb{R}^2$

10.2.1 Parametric Line Equations

Given two parametric lines in $\mathbb{R}^2$, there are two possible orientation scenarios: they intersect, or they are parallel. We begin by assuming that they intersect as shown in Fig. 10.1, with the following points and vectors:

- $R(x_r, y_r)$ is a point on the first line,
- $\mathbf{r} = x_r\mathbf{i} + y_r\mathbf{j}$ is R's position vector,
- $\mathbf{a} = x_a\mathbf{i} + y_a\mathbf{j}$ is the first line's direction vector,
- $S(x_s, y_s)$ is a point on the second line,
- $\mathbf{s} = x_s\mathbf{i} + y_s\mathbf{j}$ is S's position vector,
- $\mathbf{b} = x_b\mathbf{i} + y_b\mathbf{j}$ is the second line's direction vector,
- $P(x, y)$ is the point of intersection,
- $\mathbf{p} = x\mathbf{i} + y\mathbf{j}$ is P's position vector,
- $\lambda\mathbf{a} = \overrightarrow{RP}$,
- $\epsilon\mathbf{b} = \overrightarrow{SP}$.

We begin by defining $\mathbf{p}$ for the two lines:

$$\mathbf{p} = \mathbf{r} + \lambda\mathbf{a}$$
$$\mathbf{p} = \mathbf{s} + \epsilon\mathbf{b}.$$

Fig. 10.1 Two intersecting parametric lines

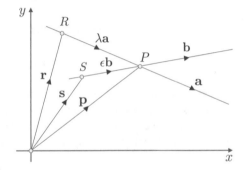

Therefore, for an intersection:

$$\mathbf{r} + \lambda\mathbf{a} = \mathbf{s} + \epsilon\mathbf{b}. \tag{10.1}$$

To eliminate ϵ in (10.1), multiply by $\mathbf{b}^{\perp}$ using the dot product:

$$\mathbf{b}^{\perp} \cdot \mathbf{r} + \lambda\mathbf{b}^{\perp} \cdot \mathbf{a} = \mathbf{b}^{\perp} \cdot \mathbf{s} + \epsilon\mathbf{b}^{\perp} \cdot \mathbf{b}.$$

But $\mathbf{b}^{\perp} \cdot \mathbf{b} = 0$, therefore,

$$\mathbf{b}^{\perp} \cdot \mathbf{r} + \lambda\mathbf{b}^{\perp} \cdot \mathbf{a} = \mathbf{b}^{\perp} \cdot \mathbf{s}. \tag{10.2}$$

Rearrange (10.2),

$$\lambda = \frac{\mathbf{b}^{\perp} \cdot (\mathbf{s} - \mathbf{r})}{\mathbf{b}^{\perp} \cdot \mathbf{a}} \tag{10.3}$$

where,

$$\mathbf{b}^{\perp} = -y_b\mathbf{i} + x_b\mathbf{j}. \tag{10.4}$$

Substitute (10.4) in (10.3):

$$
\begin{aligned}
\lambda &= \frac{(-y_b\mathbf{i} + x_b\mathbf{j}) \cdot ((x_s - x_r)\mathbf{i} + (y_s - y_r)\mathbf{j})}{(-y_b\mathbf{i} + x_b\mathbf{j}) \cdot (x_a\mathbf{i} + y_a\mathbf{j})} \\
&= \frac{x_b(y_s - y_r) - y_b(x_s - x_r)}{x_b y_a - x_a y_b}.
\end{aligned} \tag{10.5}
$$

(10.5) can be written using determinants:

$$\lambda = \frac{\begin{vmatrix} x_b & y_b \\ x_s - x_r & y_s - y_r \end{vmatrix}}{\begin{vmatrix} x_b & y_b \\ x_a & y_a \end{vmatrix}}. \tag{10.6}$$

To eliminate λ from (10.1), multiply by $\mathbf{a}^{\perp}$ using the dot product:

$$\mathbf{a}^{\perp} \cdot \mathbf{r} + \lambda\mathbf{a}^{\perp} \cdot \mathbf{a} = \mathbf{a}^{\perp} \cdot \mathbf{s} + \epsilon\mathbf{a}^{\perp} \cdot \mathbf{b}.$$

But $\mathbf{a}^{\perp} \cdot \mathbf{a} = 0$, therefore,

$$\mathbf{a}^{\perp} \cdot \mathbf{r} = \mathbf{a}^{\perp} \cdot \mathbf{s} + \epsilon\mathbf{a}^{\perp} \cdot \mathbf{b}. \tag{10.7}$$

Rearrange (10.7),

$$\epsilon = \frac{\mathbf{a}^{\perp} \cdot (\mathbf{r} - \mathbf{s})}{\mathbf{a}^{\perp} \cdot \mathbf{b}} \tag{10.8}$$

where,

$$\mathbf{a}^\perp = -y_a\mathbf{i} + x_a\mathbf{j}. \tag{10.9}$$

Substitute (10.9) in (10.8):

$$\epsilon = \frac{(-y_a\mathbf{i} + x_a\mathbf{j}) \cdot ((x_r - x_s)\mathbf{i} + (y_r - y_s)\mathbf{j})}{(-y_a\mathbf{i} + x_a\mathbf{j}) \cdot (x_b\mathbf{i} + y_b\mathbf{j})}$$
$$= \frac{x_a(y_s - y_r) - y_a(x_s - x_r)}{x_b y_a - x_a y_b}. \tag{10.10}$$

(10.10) can be written using determinants:

$$\epsilon = \frac{\begin{vmatrix} x_a & y_a \\ x_s - x_r & y_s - y_r \end{vmatrix}}{\begin{vmatrix} x_b & y_b \\ x_a & y_a \end{vmatrix}}. \tag{10.11}$$

The coordinates of P are:

$$x = x_r + \lambda x_a, \quad y = y_r + \lambda y_a$$

or

$$x = x_s + \epsilon x_b, \quad y = y_s + \epsilon y_b.$$

The lines are parallel if

$$\begin{vmatrix} x_b & y_b \\ x_a & y_a \end{vmatrix} = 0.$$

Let's test the above equations with the example shown in Fig. 10.2, with the following points and vectors:

- $R(0, 0)$ is a point on the first line,
- $\mathbf{r} = 0\mathbf{i} + 0\mathbf{j}$ is R's position vector,
- $\mathbf{a} = 1\mathbf{i} + 1\mathbf{j}$ is the first line's direction vector,
- $S(0, 1)$ is a point on the second line,
- $\mathbf{s} = 0\mathbf{i} + 1\mathbf{j}$ is S's position vector,
- $\mathbf{b} = 1\mathbf{i} - 1\mathbf{j}$ is the second line's direction vector,
- $P(x, y)$ is the point of intersection,
- $\mathbf{p} = x\mathbf{i} + y\mathbf{j}$ is P's position vector,
- $\lambda\mathbf{a} = \overrightarrow{RP}$.

Fig. 10.2 Two intersecting
parametric lines

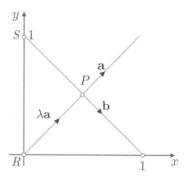

Therefore, using (10.6)

$$\lambda = \frac{\begin{vmatrix} x_b & y_b \\ x_s - x_r & y_s - y_r \end{vmatrix}}{\begin{vmatrix} x_b & y_b \\ x_a & y_a \end{vmatrix}} = \frac{\begin{vmatrix} 1 & -1 \\ 0 & 1 \end{vmatrix}}{\begin{vmatrix} 1 & -1 \\ 1 & 1 \end{vmatrix}} = \frac{1}{1+1} = \frac{1}{2}$$

and,

$$x = 0 + \tfrac{1}{2} \times 1 = \tfrac{1}{2}, \quad y = 0 + \tfrac{1}{2} \times 1 = \tfrac{1}{2},$$

which is correct.

10.2.2 Cartesian Line Equations

In Chap. 2 we investigated the solution to two linear equations in $\mathbb{R}^2$, which we
now use to calculate the possible intersection of two lines. This technique does not
require the use of vector analysis, but is included as a comparison. Starting with two
Cartesian line equations:

$$a_1 x + b_1 y = d_1$$
$$a_2 x + b_2 y = d_2,$$

which is expressed using matrices as

$$\begin{bmatrix} a_1 & b_1 \\ a_2 & b_2 \end{bmatrix} \begin{bmatrix} x \\ y \end{bmatrix} = \begin{bmatrix} d_1 \\ d_2 \end{bmatrix}. \tag{10.12}$$

Define **A** in (10.12):

$$\mathbf{A} = \begin{bmatrix} a_1 & b_1 \\ a_2 & b_2 \end{bmatrix}.$$

Therefore,

$$\mathbf{A}^{-1} = \frac{1}{\det \mathbf{A}} \begin{bmatrix} b_2 & -b_1 \\ -a_2 & a_1 \end{bmatrix}$$

where,

$$\det \mathbf{A} = \begin{vmatrix} a_1 & b_1 \\ a_2 & b_2 \end{vmatrix} = a_1 b_2 - a_2 b_1.$$

Multiply (10.12) by $\mathbf{A}^{-1}$:

$$\frac{1}{\det \mathbf{A}} \begin{bmatrix} b_2 & -b_1 \\ -a_2 & a_1 \end{bmatrix} \begin{bmatrix} a_1 & b_1 \\ a_2 & b_2 \end{bmatrix} \begin{bmatrix} x \\ y \end{bmatrix} = \frac{1}{\det \mathbf{A}} \begin{bmatrix} b_2 & -b_1 \\ -a_2 & a_1 \end{bmatrix} \begin{bmatrix} d_1 \\ d_2 \end{bmatrix}$$

$$\begin{bmatrix} x \\ y \end{bmatrix} = \frac{1}{\det \mathbf{A}} \begin{bmatrix} b_2 & -b_1 \\ -a_2 & a_1 \end{bmatrix} \begin{bmatrix} d_1 \\ d_2 \end{bmatrix}. \qquad (10.13)$$

(10.13) provides the coordinates of an intersection, so long as $\det \mathbf{A} \neq 0$.

Let's test (10.13) using the two lines shown in Fig. 10.3. The two line equations are:

$$x + y = 1$$
$$x - y = 0$$

which in matrix notation are:

$$\begin{bmatrix} 1 & 1 \\ 1 & -1 \end{bmatrix} \begin{bmatrix} x \\ y \end{bmatrix} = \begin{bmatrix} 1 \\ 0 \end{bmatrix}. \qquad (10.14)$$

Let

$$\mathbf{A} = \begin{bmatrix} 1 & 1 \\ 1 & -1 \end{bmatrix}.$$

Therefore,

$$\det \mathbf{A} = \begin{vmatrix} 1 & 1 \\ 1 & -1 \end{vmatrix} = -2.$$

Next,

$$\mathbf{A}^{-1} = -\frac{1}{2} \begin{bmatrix} -1 & -1 \\ -1 & 1 \end{bmatrix}.$$

Therefore,

$$\begin{bmatrix} x \\ y \end{bmatrix} = -\frac{1}{2} \begin{bmatrix} -1 & -1 \\ -1 & 1 \end{bmatrix} \begin{bmatrix} 1 \\ 0 \end{bmatrix} = \begin{bmatrix} \frac{1}{2} \\ \frac{1}{2} \end{bmatrix},$$

which is correct.

Fig. 10.3 Two intersecting
Cartesian lines

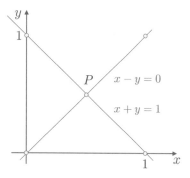

10.3   A Line Intersecting a Circle in $\mathbb{R}^2$

Now let's consider the case of a line intersecting a circle in $\mathbb{R}^2$, for which there
are three possible scenarios: the line intersects the circle at two points; the line is
tangential to the circle at a single point; or the line does not intersect the circle at all.

Figure 10.4 shows a line intersecting a circle with the following points and vectors:

- $P(x_p, y_p)$ is a point on the circle,
- $\mathbf{p} = x_p\mathbf{i} + y_p\mathbf{j}$ is P's position vector,
- $T(x_t, y_t)$ is a point on the line,
- $\mathbf{t} = x_t\mathbf{i} + y_t\mathbf{j}$ is T's position vector,
- $\hat{\mathbf{v}} = x_v\mathbf{i} + y_v\mathbf{j}$ is the line's unit direction vector,
- $\lambda\hat{\mathbf{v}} = \overrightarrow{TP}$,
- r is the circle's radius.

The algebraic equation of a circle centered at the origin is

$$x_p^2 + y_p^2 = r^2. \tag{10.15}$$

Fig. 10.4 A line intersecting
a circle

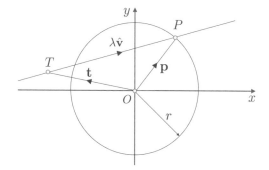

But (10.15) can also be expressed in vector form as

$$\mathbf{p} \cdot \mathbf{p} = r^2 \qquad\qquad (10.16)$$

where, from Fig. 10.4, we see that the equation for a line is

$$\mathbf{p} = \mathbf{t} + \lambda \hat{\mathbf{v}}. \qquad\qquad (10.17)$$

Therefore, for an intersection:

$$(\mathbf{t} + \lambda \hat{\mathbf{v}}) \cdot (\mathbf{t} + \lambda \hat{\mathbf{v}}) = r^2$$
$$\mathbf{t} \cdot \mathbf{t} + \lambda 2\hat{\mathbf{v}} \cdot \mathbf{t} + \lambda^2 \hat{\mathbf{v}} \cdot \hat{\mathbf{v}} = r^2. \qquad\qquad (10.18)$$

But $\hat{\mathbf{v}} \cdot \hat{\mathbf{v}} = 1$, therefore (10.18) becomes:

$$\lambda^2 + \lambda 2\hat{\mathbf{v}} \cdot \mathbf{t} + \|\mathbf{t}\|^2 - r^2 = 0. \qquad\qquad (10.19)$$

(10.19) is a quadratic in λ, and solved using

$$\lambda = \frac{-B \pm \sqrt{B^2 - 4AC}}{2A} \qquad\qquad (10.20)$$

where,

$$A = 1, \quad B = 2\mathbf{t} \cdot \hat{\mathbf{v}}, \quad C = \|\mathbf{t}\|^2 - r^2.$$

Substitute A, B, C in (10.20):

$$\lambda = \frac{-2\mathbf{t} \cdot \hat{\mathbf{v}} \pm \sqrt{4(\mathbf{t} \cdot \hat{\mathbf{v}})^2 - 4(\|\mathbf{t}\|^2 - r^2)}}{2}$$
$$= -\mathbf{t} \cdot \hat{\mathbf{v}} \pm \sqrt{(\mathbf{t} \cdot \hat{\mathbf{v}})^2 - (\|\mathbf{t}\|^2 - r^2)}$$
$$= -(x_t x_v + y_t y_v) \pm \sqrt{(x_t x_v + y_t y_v)(x_t x_v + y_t y_v) - x_t^2 - y_t^2 + r^2}$$
$$= -(x_t x_v + y_t y_v) \pm \sqrt{x_t^2 x_v^2 + 2x_t x_v y_t y_v + y_t^2 y_v^2 - x_t^2 - y_t^2 + r^2}$$
$$= -(x_t x_v + y_t y_v) \pm \sqrt{x_t^2(x_v^2 - 1) + 2x_t x_v y_t y_v + y_t^2(y_v^2 - 1) + r^2}.$$

But $x_v^2 - 1 = -y_v^2$, and $y_v^2 - 1 = -x_v^2$:

$$\lambda = -(x_t x_v + y_t y_v) \pm \sqrt{-x_t^2 y_v^2 + 2 x_t y_t x_v y_v - y_t^2 x_v^2 + r^2}$$

$$= -(x_t x_v + y_t y_v) \pm \sqrt{r^2 - \left(x_t^2 y_v^2 - 2 x_t y_t x_v y_v + y_t^2 x_v^2\right)}$$

$$= -(x_t x_v + y_t y_v) \pm \sqrt{r^2 - \left(x_t y_v - y_t x_v\right)^2}$$

$$\lambda = -(x_t x_v + y_t y_v) \pm \sqrt{r^2 - \begin{vmatrix} x_t & y_t \\ x_v & y_v \end{vmatrix}^2}. \tag{10.21}$$

The value of the discriminant of (10.21) determines whether the line misses, touches or intersects the circle:

Miss condition:

$$r^2 - \begin{vmatrix} x_t & y_t \\ x_v & y_v \end{vmatrix}^2 < 0.$$

Touch condition:

$$r^2 - \begin{vmatrix} x_t & y_t \\ x_v & y_v \end{vmatrix}^2 = 0.$$

Intersect condition:

$$r^2 - \begin{vmatrix} x_t & y_t \\ x_v & y_v \end{vmatrix}^2 > 0.$$

Let's test these conditions using the three lines and circle shown in Fig. 10.5.

Figure 10.5 shows three lines that miss, touch and intersect a circle with the following points and vectors:

- $r = 1$ is the circle's radius,
- L_1 is a line that misses the circle,
- $\hat{\mathbf{v}}_1 = 0\mathbf{i} + 1\mathbf{j}$ is L_1's unit direction vector,
- $T_1(2, 0)$ is a point on L_1,

Fig. 10.5 Three lines and a circle

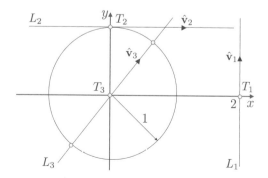

- $\mathbf{t}_1 = 2\mathbf{i} + 0\mathbf{j}$ is T_1's position vector,
- L_2 is a line that touches the circle,
- $\hat{\mathbf{v}}_2 = 0\mathbf{i} + 1\mathbf{j}$ is L_2's unit direction vector,
- $T_2(0, 1)$ is a point on L_2,
- $\mathbf{t}_2 = 0\mathbf{i} + 1\mathbf{j}$ is T_2's position vector,
- L_3 is a line that intersects the circle,
- $\hat{\mathbf{v}}_3 = \frac{1}{\sqrt{2}}\mathbf{i} + \frac{1}{\sqrt{2}}\mathbf{j}$ is L_3's unit direction vector,
- $T_3(0, 0)$ is a point on L_3,
- $\mathbf{t}_3 = 0\mathbf{i} + 0\mathbf{j}$ is T_3's position vector.

From Fig. 10.5, L_1 misses the circle, therefore,

$$r^2 - \begin{vmatrix} x_t & y_t \\ x_v & y_v \end{vmatrix}^2 = 1^2 - \begin{vmatrix} 2 & 0 \\ 0 & 1 \end{vmatrix}^2 = -3.$$

From Fig. 10.5, L_2 touches the circle, therefore,

$$r^2 - \begin{vmatrix} x_t & y_t \\ x_v & y_v \end{vmatrix}^2 = 1^2 - \begin{vmatrix} 0 & 1 \\ 1 & 0 \end{vmatrix}^2 = 0.$$

Therefore,

$$\lambda = -(x_t x_v + y_t y_v) = -(0 \times 1 + 1 \times 0) = 0$$

and

$$\mathbf{p} = \mathbf{t}_2 + \lambda\hat{\mathbf{v}}_2 = (0\mathbf{i} + 1\mathbf{j}) + 0(1\mathbf{i} + 0\mathbf{j}) = 0\mathbf{i} + 1\mathbf{j}.$$

L_2 touches the circle at $(0, 1)$.

From Fig. 10.5, L_3 intersects the circle, therefore,

$$r^2 - \begin{vmatrix} x_t & y_t \\ x_v & y_v \end{vmatrix}^2 = 1^2 - \begin{vmatrix} 0 & 0 \\ \frac{1}{\sqrt{2}} & \frac{1}{\sqrt{2}} \end{vmatrix}^2 = 1.$$

Therefore,

$$\lambda = -(x_t x_v + y_t y_v) \pm 1 = -\left(0 \times \tfrac{1}{\sqrt{2}} + 0 \times \tfrac{1}{\sqrt{2}}\right) \pm 1 = \pm 1.$$

With $\lambda = +1$:

$$\mathbf{p} = \mathbf{t}_3 + \lambda\hat{\mathbf{v}}_3$$
$$= (0\mathbf{i} + 0\mathbf{j}) + 1\left(\tfrac{1}{\sqrt{2}}\mathbf{i} + \tfrac{1}{\sqrt{2}}\mathbf{j}\right)$$
$$(x_{p1}, y_{p1}) = \left(\tfrac{1}{\sqrt{2}}, \tfrac{1}{\sqrt{2}}\right).$$

With $\lambda = -1$:

$$\mathbf{p} = \mathbf{t}_3 + \lambda \hat{\mathbf{v}}_3$$

$$= (0\mathbf{i} + 0\mathbf{j}) - 1\left(\tfrac{1}{\sqrt{2}}\mathbf{i} + \tfrac{1}{\sqrt{2}}\mathbf{j}\right)$$

$$(x_{p2}, y_{p2}) = \left(-\tfrac{1}{\sqrt{2}}, -\tfrac{1}{\sqrt{2}}\right).$$

L_3 intersects the circle at $\left(\tfrac{1}{\sqrt{2}}, \tfrac{1}{\sqrt{2}}\right)$, and $\left(-\tfrac{1}{\sqrt{2}}, -\tfrac{1}{\sqrt{2}}\right)$.

10.4 A Line Intersecting an Ellipse in $\mathbb{R}^2$

In this section we develop the previous idea of the line-circle intersection with a line-ellipse intersection. We begin with the equation of an ellipse centered at the origin

$$\frac{x_p^2}{a^2} + \frac{y_p^2}{b^2} = 1 \qquad (10.22)$$

where a and b are the x- and y-radius, respectively.

Figure 10.6 shows a line intersecting an ellipse with the following points and vectors:

- $P(x_p, y_p)$ is a point on the ellipse,
- $\mathbf{p} = x_p\mathbf{i} + y_p\mathbf{j}$ is P's position vector,
- $T(x_t, y_t)$ is a point on the line,
- $\mathbf{t} = x_t\mathbf{i} + y_t\mathbf{j}$ is T's position vector,
- $\hat{\mathbf{v}} = x_v\mathbf{i} + y_v\mathbf{j}$ is the line's unit direction vector,
- $\lambda\hat{\mathbf{v}} = \overrightarrow{TP}$,
- a is the ellipse's x-radius,
- b is the ellipse's y-radius.

Fig. 10.6 A line intersecting an ellipse

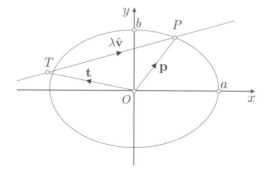

From Fig. 10.6 we see that:

$$\mathbf{p} = \mathbf{t} + \lambda \hat{\mathbf{v}} \tag{10.23}$$

where,

$$x_p = x_t + \lambda x_v \tag{10.24}$$
$$y_p = y_t + \lambda y_v. \tag{10.25}$$

Next, we substitute (10.24) and (10.25) in (10.22):

$$\frac{(x_t + \lambda x_v)}{a^2} + \frac{(y_t + \lambda y_v)}{a^2} = 1$$
$$b^2(x_t^2 + 2\lambda x_t x_v + \lambda^2 x_v^2) + a^2(y_t^2 + 2\lambda y_t y_v + \lambda^2 y_v^2) = 1$$
$$\lambda^2(b^2 x_v^2 + a^2 y_v^2) + \lambda(2b^2 x_t x_v + 2a^2 y_t y_v) + a^2 y_t^2 + b^2 x_t^2 - a^2 b^2 = 0. \tag{10.26}$$

(10.26) is a quadratic in λ, and solved using:

$$\lambda = \frac{-B \pm \sqrt{B^2 - 4AC}}{2A} \tag{10.27}$$

where,

$$A = b^2 x_v^2 + a^2 y_v^2, \quad B = 2b^2 x_t x_v + 2a^2 y_t y_v, \quad C = a^2 y_t^2 + b^2 x_t^2 - a^2 b^2.$$

Substitute A, B, C in (10.27):

$$\lambda = \frac{-(b^2 x_t x_v + a^2 y_t y_v) \pm \sqrt{(b^2 x_t x_v + a^2 y_t y_v)^2 - (b^2 x_v^2 + a^2 y_v^2)(a^2 y_t^2 + b^2 x_t^2 - a^2 b^2)}}{b^2 x_v^2 + a^2 y_v^2}$$

$$= \frac{-\left(\dfrac{x_t x_v}{a^2} + \dfrac{y_t y_v}{b^2}\right) \pm \sqrt{\dfrac{x_v^2}{a^2} + \dfrac{y_v^2}{b^2} - \dfrac{\begin{vmatrix} x_t & y_t \\ x_v & y_v \end{vmatrix}^2}{a^2 b^2}}}{\dfrac{x_v^2}{a^2} + \dfrac{y_v^2}{b^2}}. \tag{10.28}$$

Once again, the value of the discriminant reflects whether the line misses, touches or intersects the ellipse:

Miss condition:

$$\frac{x_v^2}{a^2} + \frac{y_v^2}{b^2} - \frac{\begin{vmatrix} x_t & y_t \\ x_v & y_v \end{vmatrix}^2}{a^2 b^2} < 0.$$

Touch condition:

$$\frac{x_v^2}{a^2} + \frac{y_v^2}{b^2} - \frac{\begin{vmatrix} x_t & y_t \\ x_v & y_v \end{vmatrix}^2}{a^2 b^2} = 0.$$

Intersect condition:

$$\frac{x_v^2}{a^2} + \frac{y_v^2}{b^2} - \frac{\begin{vmatrix} x_t & y_t \\ x_v & y_v \end{vmatrix}^2}{a^2 b^2} > 0.$$

We can check the validity of (10.28) by setting $a = b = r$, which reveals:

$$\lambda = \frac{-\left(\dfrac{x_t x_v}{r^2} + \dfrac{y_t y_v}{r^2}\right) \pm \sqrt{\dfrac{x_v^2}{r^2} + \dfrac{y_v^2}{r^2} - \dfrac{\begin{vmatrix} x_t & y_t \\ x_v & y_v \end{vmatrix}^2}{r^2 r^2}}}{\dfrac{x_v^2}{r^2} + \dfrac{y_v^2}{r^2}}$$

$$= -(x_t x_v + y_t y_v) \pm \sqrt{r^2 - \begin{vmatrix} x_t & y_t \\ x_v & y_v \end{vmatrix}^2} \qquad (10.29)$$

and λ in (10.29) is the same as λ in (10.25), for the circle.

Now let's test (10.28) with three lines that miss, touch and intersect an ellipse.

Figure 10.7 shows three lines that miss, touch and intersect an ellipse with the following points and vectors:

- $a = 1$ is the ellipse's horizontal radius,
- $b = 1$ is the ellipse's vertical radius,
- L_1 is a line that misses the ellipse,
- $\hat{\mathbf{v}}_1 = 0\mathbf{i} + 1\mathbf{j}$ is L_1's unit direction vector,
- $T_1(2, 0)$ is a point on L_1,
- $\mathbf{t}_1 = 2\mathbf{i} + 0\mathbf{j}$ is T_1's position vector,

Fig. 10.7 Three lines and an ellipse

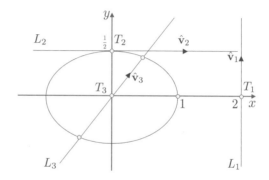

- L_2 is a line that touches the ellipse,
- $\hat{v}_2 = 0\mathbf{i} + 1\mathbf{j}$ is L_2's unit direction vector,
- $T_2(0, \frac{1}{2})$ is a point on L_2,
- $\mathbf{t}_2 = 0\mathbf{i} + \frac{1}{2}\mathbf{j}$ is T_2's position vector,
- L_3 is a line that intersects the ellipse,
- $\hat{v}_3 = \frac{1}{\sqrt{2}}\mathbf{i} + \frac{1}{\sqrt{2}}\mathbf{j}$ is L_3's unit direction vector,
- $T_3(0, 0)$ is a point on L_3,
- $\mathbf{t}_3 = 0\mathbf{i} + 0\mathbf{j}$ is T_3's position vector.

From Fig. 10.7, L_1 misses the ellipse, therefore,

$$\frac{x_v^2}{a^2} + \frac{y_v^2}{b^2} - \frac{\begin{vmatrix} x_t & y_t \\ x_v & y_v \end{vmatrix}^2}{a^2 b^2} = \frac{0^2}{1^2} + \frac{1^2}{\frac{1}{2}^2} - \frac{\begin{vmatrix} 0 & 1 \\ 2 & 0 \end{vmatrix}^2}{1^2 \frac{1}{2}^2} = 4 - 16 = -12.$$

From Fig. 10.7, L_2 touches the ellipse, therefore,

$$\frac{x_v^2}{a^2} + \frac{y_v^2}{b^2} - \frac{\begin{vmatrix} x_t & y_t \\ x_v & y_v \end{vmatrix}^2}{a^2 b^2} = \frac{1^2}{1^2} + \frac{0^2}{\frac{1}{2}^2} - \frac{\begin{vmatrix} 0 & \frac{1}{2} \\ 1 & 0 \end{vmatrix}^2}{1^2 \frac{1}{2}^2} = 1 - 1 = 0.$$

Therefore,

$$\lambda = \frac{-\left(\dfrac{x_t x_v}{a^2} + \dfrac{y_t y_v}{b^2}\right)}{\dfrac{x_v^2}{a^2} + \dfrac{y_v^2}{b^2}} = \frac{-\left(\dfrac{0 \times 1}{1^2} + \dfrac{\frac{1}{2} \times 0}{\frac{1}{2}^2}\right)}{\dfrac{1^2}{1^2} + \dfrac{0^2}{\frac{1}{2}^2}} = 0$$

and

$$\mathbf{p} = \mathbf{t} + \lambda \hat{v}_2$$
$$= \mathbf{t} + 0\hat{v}_2$$
$$= \mathbf{t}$$
$$(x_p, y_p) = \left(0, \frac{1}{2}\right).$$

L_2 touches the ellipse at $\left(0, \frac{1}{2}\right)$.

From Fig. 10.7, L_3 intersects the ellipse, therefore,

$$\frac{x_v^2}{a^2} + \frac{y_v^2}{b^2} - \frac{\begin{vmatrix} x_t & y_t \\ x_v & y_v \end{vmatrix}^2}{a^2 b^2} = \frac{\frac{1}{\sqrt{2}}^2}{1^2} + \frac{\frac{1}{\sqrt{2}}^2}{\frac{1}{2}^2} - \frac{\begin{vmatrix} 0 & 0 \\ \frac{1}{\sqrt{2}} & \frac{1}{\sqrt{2}} \end{vmatrix}^2}{1^2 \frac{1}{2}^2} = \frac{1}{2} + 2 = 2.5.$$

Therefore,

$$\lambda = \frac{-\left(\dfrac{0 \times \frac{1}{\sqrt{2}}}{1^2} + \dfrac{0 \times \frac{1}{\sqrt{2}}}{\frac{1}{2}^2}\right) \pm \sqrt{2.5}}{\dfrac{\frac{1}{\sqrt{2}}^2}{1^2} + \dfrac{\frac{1}{\sqrt{2}}^2}{\frac{1}{2}^2}} = \frac{\sqrt{2.5}}{\frac{1}{2} + 2} = \pm\frac{\sqrt{2.5}}{2.5}$$

and

$$\mathbf{p} = \mathbf{t} \pm \lambda\hat{\mathbf{v}}_3$$
$$= \pm\lambda\hat{\mathbf{v}}_3$$
$$= \pm\frac{\sqrt{2.5}}{2.5}\left(\tfrac{1}{\sqrt{2}}\mathbf{i} + \tfrac{1}{\sqrt{2}}\mathbf{j}\right)$$
$$\approx \pm(0.447\mathbf{i} + 0.447\mathbf{j})$$
$$(x_p, y_p) \approx \pm(0.447, 0.447).$$

L_3 intersects the ellipse at $\approx (0.447, 0.447)$ and $\approx (-0.447, -0.447)$.

10.5   The Shortest Distance Between Two Skew Lines in $\mathbb{R}^3$

Having seen how to cope with two-dimensional lines, circles and ellipses, now let's explore the relationship between two lines in $\mathbb{R}^3$.

We already know that lines in $\mathbb{R}^2$ either intersect or are parallel with one another. However, in $\mathbb{R}^3$ a third option is possible—one where the lines approach one another then recede, allowing a shortest distance to be calculated. Such lines are called *skew* lines, and the shortest distance between them will be the length of a mutual perpendicular to both lines.

Figure 10.8 shows two skew lines with the following points and vectors:

- L_1 is the first skew line,
- $P_1(x_1, y_1, z_1)$ is a point on L_1 at the closest point,
- $\mathbf{p}_1 = x_1\mathbf{i} + y_1\mathbf{j} + z_1\mathbf{k}$ is P_1's position vector,
- $T_1(x_{t1}, y_{t1}, z_{t1})$ is a point on L_1,
- $\mathbf{t}_1 = x_{t1}\mathbf{i} + y_{t1}\mathbf{j} + z_{t1}\mathbf{k}$ is T_1's position vector,
- $\mathbf{v}_1 = x_{v1}\mathbf{i} + y_{v1}\mathbf{j} + z_{v1}\mathbf{k}$ is L_1's direction vector,
- $\lambda\mathbf{v}_1 = \overrightarrow{T_1 P_1}$,
- L_2 is the second skew line,
- $P_2(x_2, y_2, z_2)$ is a point on L_2 at the closest point,
- $\mathbf{p}_2 = x_2\mathbf{i} + y_2\mathbf{j} + z_2\mathbf{k}$ is P_2's position vector,
- $T_2(x_{t2}, y_{t2}, z_{t2})$ is a point on L_2,

Fig. 10.8 Two skew lines

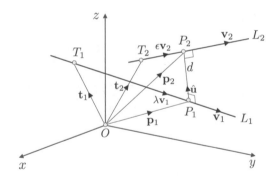

- $\mathbf{t}_2 = x_{t2}\mathbf{i} + y_{t2}\mathbf{j} + z_{t2}\mathbf{k}$ is T_2's position vector,
- $\mathbf{v}_2 = x_{v2}\mathbf{i} + y_{v2}\mathbf{j} + z_{v2}\mathbf{k}$ is L_2's direction vector,
- $\epsilon \mathbf{v}_2 = \overrightarrow{T_2 P_2}$,
- $\hat{\mathbf{u}}$ is a unit vector parallel to $\mathbf{p}_2 - \mathbf{p}_1$.

In Fig. 10.8, the shortest distance d between the skew lines is the magnitude of the vector $\overrightarrow{P_1 P_2}$, which is perpendicular to both lines. The equations for the two lines are

$$\mathbf{p}_1 = \mathbf{t}_1 + \lambda \mathbf{v}_1$$
$$\mathbf{p}_2 = \mathbf{t}_2 + \epsilon \mathbf{v}_2.$$

Let $\hat{\mathbf{u}}$ be the unit vector parallel to $\overrightarrow{P_1 P_2}$, then

$$\hat{\mathbf{u}} = \frac{\mathbf{v}_1 \times \mathbf{v}_2}{\|\mathbf{v}_1 \times \mathbf{v}_2\|}. \tag{10.30}$$

Figure 10.9 shows a right-angled triangle formed from $\hat{\mathbf{u}}$ and $\mathbf{t}_2 - \mathbf{t}_1$. Therefore,

$$\begin{aligned} d &= \|\mathbf{t}_2 - \mathbf{t}_1\| \cos\theta \\ &= \|\hat{\mathbf{u}}\| \, \|\mathbf{t}_2 - \mathbf{t}_1\| \cos\theta \\ &= |\hat{\mathbf{u}} \cdot (\mathbf{t}_2 - \mathbf{t}_1)|. \end{aligned} \tag{10.31}$$

Substitute (10.30) in (10.31), we have:

$$d = \left| \frac{(\mathbf{v}_1 \times \mathbf{v}_2) \cdot (\mathbf{t}_2 - \mathbf{t}_1)}{\|\mathbf{v}_1 \times \mathbf{v}_2\|} \right|. \tag{10.32}$$

If $d = 0$, then the lines intersect.

Let's test (10.32) with a simple example as shown in Fig. 10.10. By inspection, the shortest line between L_1 and L_2, is a perpendicular to L_1 from the origin, which is $\frac{\sqrt{2}}{2}$ in length.

Fig. 10.9 A right-angled triangle

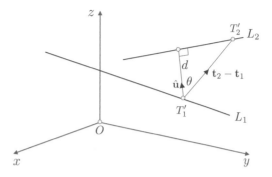

Fig. 10.10 Two skew lines

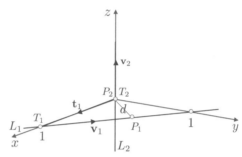

The two lines have the following definitions:

$$T_1 = (1, 0, 0)$$
$$\mathbf{t}_1 = 1\mathbf{i} + 0\mathbf{j} + 0\mathbf{k}$$
$$\mathbf{v}_1 = -1\mathbf{i} + 1\mathbf{j} + 0\mathbf{k}$$
$$T_2 = (0, 0, 0)$$
$$\mathbf{t}_2 = 0\mathbf{i} + 0\mathbf{j} + 0\mathbf{k}$$
$$\mathbf{v}_2 = 0\mathbf{i} + 0\mathbf{j} + 1\mathbf{k}.$$

Calculating $\mathbf{v}_1 \times \mathbf{v}_2$ and $\|\mathbf{v}_1 \times \mathbf{v}_2\|$:

$$\mathbf{v}_1 \times \mathbf{v}_2 = \begin{vmatrix} \mathbf{i} & \mathbf{j} & \mathbf{k} \\ -1 & 1 & 0 \\ 0 & 0 & 1 \end{vmatrix} = 1\mathbf{i} + 1\mathbf{j} + 0\mathbf{k}$$

$$\|\mathbf{v}_1 \times \mathbf{v}_2\| = \sqrt{1^2 + 1^2 + 0^2} = \sqrt{2}.$$

Therefore,

$$d = \left| \frac{(\mathbf{v}_1 \times \mathbf{v}_2) \cdot (\mathbf{t}_2 - \mathbf{t}_1)}{\|\mathbf{v}_1 \times \mathbf{v}_2\|} \right| = \left| \frac{(1\mathbf{i} + 1\mathbf{j} + 0\mathbf{k}) \cdot (-1\mathbf{i} + 0\mathbf{j} + 0\mathbf{k})}{\sqrt{2}} \right|$$

$$= \left| \frac{-1}{\sqrt{2}} \right| = \frac{\sqrt{2}}{2},$$

which confirms our prediction.

10.6   Two Intersecting Lines in $\mathbb{R}^3$

As we have seen in the previous section, two straight lines in $\mathbb{R}^3$ can cross one another, but not necessarily intersect. Therefore, in order to calculate a possible intersection, we have to ensure two things:

1. The two lines are not parallel.
2. The two lines touch.

We begin by defining the two lines as shown in Fig. 10.11, with the following points and vectors:

- L_1 is the first intersecting line,
- $P_1(x_1, y_1, z_1)$ is a point on L_1,
- $\mathbf{p}_1 = x_1\mathbf{i} + y_1\mathbf{j} + z_1\mathbf{k}$ is P_1's position vector,
- $T_1(2, 0, 1)$ is a point on L_1,
- $\mathbf{t}_1 = 2\mathbf{i} + 0\mathbf{j} + 1\mathbf{k}$ is T_1's position vector,
- $\mathbf{v}_1 = -2\mathbf{i} + 2\mathbf{j} + 1\mathbf{k}$ is L_1's direction vector,
- $\lambda\mathbf{v}_1 = \overrightarrow{T_1P_1}$,
- L_2 is the second intersecting line,
- $P_2(x_2, y_2, z_2)$ is a point on L_2,
- $\mathbf{p}_2 = x_2\mathbf{i} + y_2\mathbf{j} + z_2\mathbf{k}$ is P_2's position vector,
- $T_2(0, 2, 1)$ is a point on L_2,

Fig. 10.11   Two intersecting lines

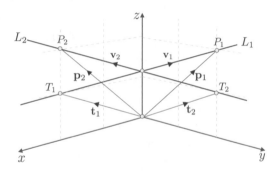

- $\mathbf{t}_2 = 0\mathbf{i} + 2\mathbf{j} + 1\mathbf{k}$ is T_2's position vector,
- $\mathbf{v}_2 = 2\mathbf{i} - 2\mathbf{j} + 1\mathbf{k}$ is L_2's direction vector,
- $\epsilon\mathbf{v}_2 = \overrightarrow{T_2 P_2}$.

Step 1: If $\mathbf{v}_1 \times \mathbf{v}_2 \neq 0$, the lines are parallel and do not intersect.
Step 2: The distance d between two skew lines is given by

$$d = \left| \frac{(\mathbf{v}_1 \times \mathbf{v}_2) \cdot (\mathbf{t}_2 - \mathbf{t}_1)}{\|\mathbf{v}_1 \times \mathbf{v}_2\|} \right|.$$

If $(\mathbf{v}_1 \times \mathbf{v}_2) \cdot (\mathbf{t}_2 - \mathbf{t}_1) \neq 0$, the lines do not intersect.

Step 3: Equate the two line equations:

$$\mathbf{t}_1 + \lambda\mathbf{v}_1 = \mathbf{t}_2 + \epsilon\mathbf{v}_2$$

$$(x_{t1}\mathbf{i} + y_{t1}\mathbf{j} + z_{t1}\mathbf{k}) + \lambda(x_{v1}\mathbf{i} + y_{v1}\mathbf{j} + z_{v1}\mathbf{k}) = (x_{t2}\mathbf{i} + y_{t2}\mathbf{j} + z_{t2}\mathbf{k}) + \epsilon(x_{v2}\mathbf{i} + y_{v2}\mathbf{j} + z_{v}\mathbf{k})$$

$$(x_{t1} - x_{t2} + \lambda x_{v1} - \epsilon x_{v2})\mathbf{i} + (y_{t1} - y_{t2} + \lambda y_{v1} - \epsilon y_{v2})\mathbf{j} + (z_{t1} - z_{t2} + \lambda z_{v1} - \epsilon z_{v2})\mathbf{k} = 0.$$

For this vector to be null, its components must vanish. Therefore, we have:

$$\lambda x_{v1} - \epsilon x_{v2} = x_{t2} - x_{t1} \tag{10.33}$$
$$\lambda y_{v1} - \epsilon y_{v2} = y_{t2} - y_{t1} \tag{10.34}$$
$$\lambda z_{v1} - \epsilon z_{v2} = z_{t2} - z_{t1} \tag{10.35}$$

which provide values for λ and ϵ which, when substituted in (10.36), reveal the intersection point. For example, the lines in Fig. 10.11 have equations:

$$\mathbf{p}_1 = \mathbf{t}_1 + \lambda\mathbf{v}_1, \quad \text{and} \quad \mathbf{p}_2 = \mathbf{t}_2 + \epsilon\mathbf{v}_2 \tag{10.36}$$

where,

$$\mathbf{t}_1 = 2\mathbf{i} + 0\mathbf{j} + 1\mathbf{k}, \quad \text{and} \quad \mathbf{t}_2 = 0\mathbf{i} + 2\mathbf{j} + 1\mathbf{k}$$

and

$$\mathbf{v}_1 = -2\mathbf{i} + 2\mathbf{j} + 1\mathbf{k}, \quad \text{and} \quad \mathbf{v}_2 = 2\mathbf{i} - 2\mathbf{j} + 1\mathbf{k}.$$

The first step is to discover whether the lines are parallel, i.e. if $\mathbf{v}_1 \times \mathbf{v}_2 = 0$.

We calculate $\mathbf{v}_1 \times \mathbf{v}_2$:

$$\mathbf{v}_1 \times \mathbf{v}_2 = \begin{vmatrix} \mathbf{i} & \mathbf{j} & \mathbf{k} \\ -2 & 2 & 1 \\ 2 & -2 & 1 \end{vmatrix} = 4\mathbf{i} + 4\mathbf{j} + 0\mathbf{k}.$$

But as $\mathbf{v}_1 \times \mathbf{v}_2 \neq 0$ the lines are not parallel.

The second step is to discover if the distance d between the two lines is zero:

$$
\begin{aligned}
d &= \left| \frac{(\mathbf{v}_1 \times \mathbf{v}_2) \cdot (\mathbf{t}_2 - \mathbf{t}_1)}{\|\mathbf{v}_1 \times \mathbf{v}_2\|} \right| \\
&= \left| \frac{(4\mathbf{i} + 4\mathbf{j} + 0\mathbf{k}) \cdot ((0\mathbf{i} + 2\mathbf{j} + 1\mathbf{k}) - (2\mathbf{i} + 0\mathbf{j} + 1\mathbf{k}))}{\|4\mathbf{i} + 4\mathbf{j} + 0\mathbf{k}\|} \right| \\
&= \left| \frac{(4\mathbf{i} + 4\mathbf{j} + 0\mathbf{k}) \cdot (-2\mathbf{i} + 2\mathbf{j} + 0\mathbf{k})}{\|\sqrt{32}\|} \right| \\
&= \left| \frac{-8 + 8 + 0}{\|\sqrt{32}\|} \right| = 0.
\end{aligned}
$$

Therefore, the lines do intersect.

The third step is to find the point of intersection using (10.33), (10.34) and (10.35):

$$
\begin{aligned}
-2\lambda - 2\epsilon &= -2 \\
2\lambda + 2\epsilon &= 2 \\
\lambda - \epsilon &= 0
\end{aligned}
$$

and

$$
\begin{aligned}
\lambda + \epsilon &= 1 \\
\lambda + \epsilon &= 1 \\
\lambda - \epsilon &= 0.
\end{aligned}
$$

Therefore,

$$\lambda = \tfrac{1}{2}, \quad \text{and} \quad \epsilon = \tfrac{1}{2}.$$

Substituting λ and ϵ in (10.36):

$$\mathbf{p}_1 = (2\mathbf{i} + 0\mathbf{j} + 1\mathbf{k}) + \tfrac{1}{2}(-2\mathbf{i} + 2\mathbf{j} + 1\mathbf{k}) = 1\mathbf{i} + 1\mathbf{j} + \tfrac{3}{2}\mathbf{k}.$$

Double checking:

$$\mathbf{p}_2 = (0\mathbf{i} + 2\mathbf{j} + 1\mathbf{k}) + \tfrac{1}{2}(2\mathbf{i} - 2\mathbf{j} + 1\mathbf{k}) = 1\mathbf{i} + 1\mathbf{j} + \tfrac{3}{2}\mathbf{k}.$$

The point of intersection is $(1, 1, 1\tfrac{1}{2})$, which is correct.

10.7   A Line Intersecting a Plane

There are two scenarios for a line and plane: they either intersect or they are parallel, and we must be able to detect both possibilities. The plane could be defined either using a plane equation or using two vectors, and as the former is the most probable format let's proceed with this.

We define the plane equation using the Cartesian form:

$$ax + by + cz = d$$

where the normal vector $\mathbf{n}$ is given by

$$\mathbf{n} = a\mathbf{i} + b\mathbf{j} + c\mathbf{k}.$$

We identify a point $P(x, y, z)$ on the plane using a position vector $\mathbf{p} = x\mathbf{i} + y\mathbf{j} + z\mathbf{k}$. And as we have seen in previous chapters, we can state:

$$\mathbf{n} \cdot \mathbf{p} = d \tag{10.37}$$

which reminds us that if $\mathbf{n}$ is a unit vector, the projection of $\mathbf{p}$ on $\mathbf{n}$ equals the perpendicular distance d from the origin to the plane.

Figure 10.12 shows the scenario described above with the following points and vectors:

- $P(x, y, z)$ is a point on the plane,
- $\mathbf{p} = x\mathbf{i} + y\mathbf{j} + z\mathbf{k}$ is P's position vector,
- $\mathbf{n} = a\mathbf{i} + b\mathbf{j} + c\mathbf{k}$ is the plane's normal vector,
- $T(x_t, y_t, z_t)$ is a point on the line,
- $\mathbf{t} = x_t\mathbf{i} + y_t\mathbf{j} + z_t\mathbf{k}$ is T's position vector,
- $\mathbf{v} = x_v\mathbf{i} + y_v\mathbf{j} + z_v\mathbf{k}$ is the line's direction vector,
- $\mathbf{p} = \mathbf{t} + \lambda\mathbf{v}$ is the equation of the line.

The line and plane equations can be satisfied for some value of λ, which can be found by substituting $\mathbf{p}$ into (10.37):

Fig. 10.12 A line intersecting a plane

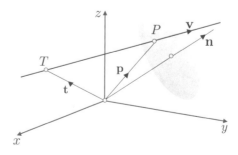

Fig. 10.13 A line
intersecting a plane

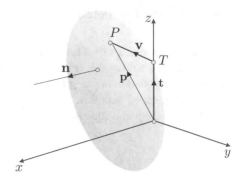

$$\mathbf{n} \cdot (\mathbf{t} + \lambda \mathbf{v}) = d$$
$$\mathbf{n} \cdot \mathbf{t} + \lambda \mathbf{n} \cdot \mathbf{v} = d$$
$$\lambda = \frac{d - \mathbf{n} \cdot \mathbf{t}}{\mathbf{n} \cdot \mathbf{v}}.$$

However, if the line is perpendicular to the planes normal, $\mathbf{n} \cdot \mathbf{v} = 0$, and λ will be undefined. Furthermore, if both $\mathbf{n}$ and $\mathbf{v}$ are unit vectors, then

$$\lambda = d - \hat{\mathbf{n}} \cdot \mathbf{t}.$$

Finally, the position vector for the point of intersection is given by

$$\mathbf{p} = \mathbf{t} + \lambda \mathbf{v}.$$

To test the technique, consider the scenario shown in Fig. 10.13, with the following points and vectors:

- $P(x, y, z)$ is a point on the plane,
- $\mathbf{p} = x\mathbf{i} + y\mathbf{j} + z\mathbf{k}$ is P's position vector,
- $x = 10$ is the plane's equation,
- $\mathbf{n} = 1\mathbf{i} + 0\mathbf{j} + 0\mathbf{k}$ is the plane's normal vector,
- $T(0, 0, 10)$ is a point on the line,
- $\mathbf{t} = 0\mathbf{i} + 0\mathbf{j} + 10\mathbf{k}$ is T's position vector,
- $\mathbf{v} = 1\mathbf{i} + 1\mathbf{j} + 1\mathbf{k}$ is the line's direction vector.

The plane equation is

$$x = 10$$

and the line equation is

$$\mathbf{p} = \mathbf{t} + \lambda \mathbf{v}.$$

Therefore,

$$\lambda = \frac{d - \mathbf{n} \cdot \mathbf{t}}{\mathbf{n} \cdot \mathbf{v}}$$
$$= \frac{10 - (1\mathbf{i} + 0\mathbf{j} + 0\mathbf{k}) \cdot (0\mathbf{i} + 0\mathbf{j} + 10\mathbf{k})}{(1\mathbf{i} + 0\mathbf{j} + 0\mathbf{k}) \cdot (1\mathbf{i} + 1\mathbf{j} + 1\mathbf{k})}$$
$$= \frac{10 - 0}{1}$$
$$= 10.$$

The point of intersection is given by:

$$\mathbf{p} = \mathbf{t} + \lambda \mathbf{v}$$
$$= (0\mathbf{i} + 0\mathbf{j} + 10\mathbf{k}) + 10(1\mathbf{i} + 1\mathbf{j} + 1\mathbf{k})$$
$$= 10\mathbf{i} + 10\mathbf{j} + 20\mathbf{k}$$
$$P = (10, 10, 20),$$

which is correct.

10.8 A Line Intersecting a Sphere

The line-sphere intersection problem is central to ray tracing and is worth investigating. However, before we start exploring the associated geometry, let's pause and remember the case of the line-circle intersection problem in Sect. 10.3.

Figure 10.4 shows a line intersecting a circle, but Fig. 10.14 is a similar diagram, but with a z-axis, and illustrates a line intersecting a sphere. Therefore, surely, the same vector analysis applies, apart from the fact that the vectors are now in $\mathbb{R}^3$ rather than $\mathbb{R}^2$?

Fig. 10.14 A line intersecting a sphere

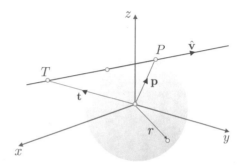

The algebraic equation of a sphere centered at the origin is

$$x_p^2 + y_p^2 + z_p^2 = r^2. \tag{10.38}$$

But (10.38) can also be expressed in vector form as

$$\mathbf{p} \cdot \mathbf{p} = r^2 \tag{10.39}$$

where, from Fig. 10.14, we see that the equation for a line is

$$\mathbf{p} = \mathbf{t} + \lambda \hat{\mathbf{v}}. \tag{10.40}$$

Therefore, for an intersection:

$$(\mathbf{t} + \lambda \hat{\mathbf{v}}) \cdot (\mathbf{t} + \lambda \hat{\mathbf{v}}) = r^2$$
$$\mathbf{t} \cdot \mathbf{t} + \lambda 2 \hat{\mathbf{v}} \cdot \mathbf{t} + \lambda^2 \hat{\mathbf{v}} \cdot \hat{\mathbf{v}} = r^2. \tag{10.41}$$

But $\hat{\mathbf{v}} \cdot \hat{\mathbf{v}} = 1$, therefore (10.41) becomes:

$$\lambda^2 + \lambda 2 \hat{\mathbf{v}} \cdot \mathbf{t} + \|\mathbf{t}\|^2 - r^2 = 0. \tag{10.42}$$

(10.42) is a quadratic in λ, and solved using

$$\lambda = \frac{-B \pm \sqrt{B^2 - 4AC}}{2A} \tag{10.43}$$

where,

$$A = 1, \quad B = 2\mathbf{t} \cdot \hat{\mathbf{v}}, \quad C = \|\mathbf{t}\|^2 - r^2.$$

Substitute A, B, C in (10.43):

$$\lambda = \frac{-2\mathbf{t} \cdot \hat{\mathbf{v}} \pm \sqrt{4(\mathbf{t} \cdot \hat{\mathbf{v}})^2 - 4(\|\mathbf{t}\|^2 - r^2)}}{2}$$
$$= -\mathbf{t} \cdot \hat{\mathbf{v}} \pm \sqrt{(\mathbf{t} \cdot \hat{\mathbf{v}})^2 - \|\mathbf{t}\|^2 + r^2}.$$

Simplifying the discriminant:

$$(\mathbf{t} \cdot \hat{\mathbf{v}})^2 - \|\mathbf{t}\|^2 + r^2 = (x_t x_v + y_t y_v + z_t z_v)(x_t x_v + y_t y_v + z_t z_v)$$
$$- x_t^2 - y_t^2 - z_t^2 + r^2 = r^2 + x_t^2 x_v^2 + x_t x_v y_t y_v$$
$$+ x_t x_v z_t z_v + y_t^2 y_v^2 + x_t x_v y_t y_v + y_t y_v z_t z_v$$
$$+ z_t^2 z_v^2 + x_t x_v z_t z_v + y_t y_v z_t z_v - x_t^2 - y_t^2 - z_t^2$$
$$= r^2 + \left(x_t^2\left(x_v^2 - 1\right) + y_t^2\left(y_v^2 - 1\right) + z_t^2\left(z_v^2 - 1\right)\right.$$
$$\left. + 2x_t x_v y_t y_v + 2x_t x_v z_t z_v + 2y_t y_v z_t z_v\right).$$

But $x_v^2 - 1 = -\left(y_v^2 + z_v^2\right),\quad y_v^2 - 1 = -\left(z_v^2 + x_v^2\right),\quad z_v^2 - 1 = -\left(x_v^2 + y_v^2\right)$:

$$(\mathbf{t} \cdot \hat{\mathbf{v}})^2 - \|\mathbf{t}\|^2 + r^2 = r^2 - \left(x_t^2\left(y_v^2 + z_v^2\right) + y_t^2\left(z_v^2 + x_v^2\right) + z_t^2\left(x_v^2 + y_v^2\right)\right.$$
$$\left. - 2x_t x_v y_t y_v - 2x_t x_v z_t z_v - 2y_t y_v z_t z_v\right)$$
$$= r^2 - \left(x_t^2 y_v^2 + x_t^2 z_v^2 + y_t^2 z_v^2 + x_v^2 y_t^2 + x_v^2 z_t^2 + y_v^2 z_t^2\right.$$
$$\left. - 2x_t x_v y_t y_v - 2x_t x_v z_t z_v - 2y_t y_v z_t z_v\right)$$
$$= r^2 - \left((x_v y_t - x_t y_v)^2 + (y_v z_t - y_t z_v)^2 + (z_v x_t - z_t x_v)^2\right)$$
$$= r^2 - \begin{vmatrix} x_v & y_v \\ x_t & y_t \end{vmatrix}^2 - \begin{vmatrix} y_v & z_v \\ y_t & z_t \end{vmatrix}^2 - \begin{vmatrix} z_v & x_v \\ z_t & x_t \end{vmatrix}^2.$$

Therefore,

$$\lambda = -(x_t x_v + y_t y_v + z_t z_v) \pm \sqrt{r^2 - \begin{vmatrix} x_v & y_v \\ x_t & y_t \end{vmatrix}^2 - \begin{vmatrix} y_v & z_v \\ y_t & z_t \end{vmatrix}^2 - \begin{vmatrix} z_v & x_v \\ z_t & x_t \end{vmatrix}^2}. \quad (10.44)$$

The value of the discriminant of (10.44) determines whether the line misses, touches or intersects the circle:

Miss condition:

$$r^2 - \begin{vmatrix} x_v & y_v \\ x_t & y_t \end{vmatrix}^2 - \begin{vmatrix} y_v & z_v \\ y_t & z_t \end{vmatrix}^2 - \begin{vmatrix} z_v & x_v \\ z_t & x_t \end{vmatrix}^2 < 0.$$

Touch condition:

$$r^2 - \begin{vmatrix} x_v & y_v \\ x_t & y_t \end{vmatrix}^2 - \begin{vmatrix} y_v & z_v \\ y_t & z_t \end{vmatrix}^2 - \begin{vmatrix} z_v & x_v \\ z_t & x_t \end{vmatrix}^2 = 0.$$

Intersect condition:

$$r^2 - \begin{vmatrix} x_v & y_v \\ x_t & y_t \end{vmatrix}^2 - \begin{vmatrix} y_v & z_v \\ y_t & z_t \end{vmatrix}^2 - \begin{vmatrix} z_v & x_v \\ z_t & x_t \end{vmatrix}^2 > 0.$$

Fig. 10.15 Three lines and a sphere

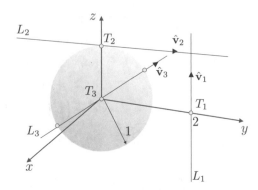

Let's test the above formulae with the unit-radius sphere shown in Fig. 10.15, with the following points and vectors:

- $r = 1$ is the sphere's radius,
- L_1 is a line that misses the sphere,
- $\hat{\mathbf{v}}_1 = 0\mathbf{i} + 0\mathbf{j} + 1\mathbf{k}$ is L_1's unit direction vector,
- $T_1(0, 2, 0)$ is a point on L_1,
- $\mathbf{t}_1 = 0\mathbf{i} + 2\mathbf{j} + 0\mathbf{k}$ is T_1's position vector,
- L_2 is a line that touches the sphere,
- $\hat{\mathbf{v}}_2 = 0\mathbf{i} + 1\mathbf{j} + 0\mathbf{k}$ is L_2's unit direction vector,
- $T_2(0, 0, 1)$ is a point on L_2,
- $\mathbf{t}_2 = 0\mathbf{i} + 0\mathbf{j} + 1\mathbf{k}$ is T_2's position vector,
- L_3 is a line that intersects the sphere,
- $\hat{\mathbf{v}}_3 = 0\mathbf{i} + \frac{1}{\sqrt{2}}\mathbf{j} + \frac{1}{\sqrt{2}}\mathbf{k}$ is L_3's unit direction vector,
- $T_3(0, 0, 0)$ is a point on L_3,
- $\mathbf{t}_3 = 0\mathbf{i} + 0\mathbf{j} + 0\mathbf{k}$ is T_3's position vector.

From Fig. 10.15, L_1 misses the sphere, therefore,

$$1^2 - \begin{vmatrix} 0 & 0 \\ 0 & 2 \end{vmatrix}^2 - \begin{vmatrix} 0 & 1 \\ 2 & 0 \end{vmatrix}^2 - \begin{vmatrix} 1 & 0 \\ 0 & 0 \end{vmatrix}^2 = -3.$$

From Fig. 10.15, L_2 touches the sphere, therefore,

$$1^2 - \begin{vmatrix} 0 & 1 \\ 0 & 0 \end{vmatrix}^2 - \begin{vmatrix} 1 & 0 \\ 0 & 1 \end{vmatrix}^2 - \begin{vmatrix} 0 & 0 \\ 1 & 0 \end{vmatrix}^2 = 0.$$

Therefore,

$$\lambda = -\mathbf{t} \cdot \hat{\mathbf{v}} = (0\mathbf{i} + 0\mathbf{j} + 1\mathbf{k}) \cdot (0\mathbf{i} + 1\mathbf{j} + 0\mathbf{k}) = 0$$

and
$$\mathbf{p}_2 = \mathbf{t}_2 = (0\mathbf{i} + 0\mathbf{j} + 1\mathbf{k}).$$

Therefore, the touch point is $(0, 0, 1)$.
From Fig. 10.15, L_3 intersects the sphere, therefore,

$$1^2 - \begin{vmatrix} 0 & \frac{1}{\sqrt{2}} \\ 0 & 0 \end{vmatrix}^2 - \begin{vmatrix} \frac{1}{\sqrt{2}} & \frac{1}{\sqrt{2}} \\ 0 & 0 \end{vmatrix}^2 - \begin{vmatrix} \frac{1}{\sqrt{2}} & 0 \\ 0 & 0 \end{vmatrix}^2 = 1.$$

Therefore,

$$\lambda = -\mathbf{t} \cdot \hat{\mathbf{v}} = (0\mathbf{i} + 0\mathbf{j} + 0\mathbf{k}) \cdot \left(0\mathbf{i} + \tfrac{1}{\sqrt{2}}\mathbf{j} + \tfrac{1}{\sqrt{2}}\mathbf{k} \right) \pm 1 = \pm 1$$

and

$$\mathbf{p}_3 = \mathbf{t}_3 + \lambda \mathbf{v}_3 = (0\mathbf{i} + 0\mathbf{j} + 0\mathbf{k}) \pm \left(0\mathbf{i} + \tfrac{1}{\sqrt{2}}\mathbf{j} + \tfrac{1}{\sqrt{2}}\mathbf{k} \right).$$

The intersecting points are $\left(0, -\tfrac{1}{\sqrt{2}}, -\tfrac{1}{\sqrt{2}} \right)$ and $\left(0, \tfrac{1}{\sqrt{2}}, \tfrac{1}{\sqrt{2}} \right)$.

10.9   A Line Intersecting a Cylinder

Another primitive used in ray tracing is the cylinder. So let's consider a cylinder with radius r and infinite length with its rotational axis aligned with the z-axis. Figure 10.16 shows this scenario with the following points and vectors:

- $P(x, y, z)$ is a point on the cylinder,
- $\mathbf{p} = x\mathbf{i} + y\mathbf{j} + z\mathbf{k}$ is P's position vector,
- $\mathbf{r}$ is the cylinder's radial vector,
- $T(x_t, y_t, z_t)$ is a point on the line,
- $\mathbf{t} = x_t\mathbf{i} + y_t\mathbf{j} + z_t\mathbf{k}$ is T's position vector,
- $\hat{\mathbf{v}} = x_v\mathbf{i} + y_v\mathbf{j} + z_v\mathbf{k}$ is the line's unit direction vector,
- $\mathbf{p} = \mathbf{t} + \lambda\hat{\mathbf{v}}$ is the equation of the line.

The line equation is given by
$$\mathbf{p} = \mathbf{t} + \lambda\hat{\mathbf{v}} \tag{10.45}$$

and the point $P(x_p, y_p, z_p)$ is the first point of intersection.
Therefore,

$$\|\mathbf{p}\|^2 = x_p^2 + y_p^2 = r^2. \tag{10.46}$$

But from (10.45), we see that

$$x_p = x_t + \lambda x_v$$
$$y_p = y_t + \lambda y_v.$$

Fig. 10.16 A line
intersecting a cylinder

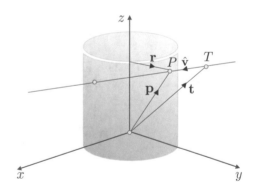

Substituting these in (10.46) we obtain

$$(x_t + \lambda x_v)^2 + (y_t + \lambda y_v)^2 = r^2. \tag{10.47}$$

Expanding (10.47) we obtain

$$\lambda^2 + \lambda 2(x_v x_t + y_v y_t) + x_t^2 + y_t^2 - r^2 = 0$$

which is a quadratic in λ, and solved using

$$\lambda = \frac{-B \pm \sqrt{B^2 - 4AC}}{2A}$$

where,

$$A = 1, \quad B = 2(x_v x_t + y_v y_t), \quad C = x_t^2 + y_t^2 - r^2$$

and can be represented by

$$\lambda = \frac{-(x_v x_t + y_v y_t) \pm \sqrt{r^2(x_v^2 + y_v^2) - \begin{vmatrix} x_v & y_v \\ x_t & y_t \end{vmatrix}^2}}{x_v^2 + y_v^2}. \tag{10.48}$$

The value of the discriminant of (10.48) determines whether the line misses, touches
or intersects the cylinder:

Miss condition:

$$r^2(x_v^2 + y_v^2) - \begin{vmatrix} x_v & y_v \\ x_t & y_t \end{vmatrix}^2 < 0.$$

Touch condition:

$$r^2(x_v^2 + y_v^2) - \begin{vmatrix} x_v & y_v \\ x_t & y_t \end{vmatrix}^2 = 0.$$

Fig. 10.17 A line
intersecting a cylinder

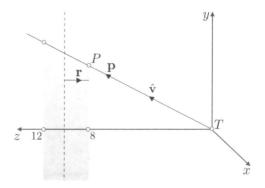

Intersect condition:

$$r^2(x_v^2 + y_v^2) - \begin{vmatrix} x_v & y_v \\ x_t & y_t \end{vmatrix}^2 > 0.$$

It is highly unlikely that the cylinder will be aligned with the z-axis, and it is simple to subject the intersecting line L to the cylinder's inverse transforms. But in order to compute the inverse transform sequence, we need to know the original transform sequence.

We start with the scenario shown in Fig. 10.17, where the cylinder has been scaled by a factor of 2, rotated $90°$ about the x-axis, and translated 10 units along the z-axis. Therefore, if L is the intersecting line in Fig. 10.17, it must be transformed as follows for the cylinder to be realigned with the z-axis:

$$L' = S^{-1}R^{-1}T^{-1}L.$$

The scale transform is:

$$S = \begin{bmatrix} 2 & 0 & 0 & 0 \\ 0 & 2 & 0 & 0 \\ 0 & 0 & 2 & 0 \\ 0 & 0 & 0 & 1 \end{bmatrix}, \quad \text{and} \quad S^{-1} = \begin{bmatrix} \frac{1}{2} & 0 & 0 & 0 \\ 0 & \frac{1}{2} & 0 & 0 \\ 0 & 0 & \frac{1}{2} & 0 \\ 0 & 0 & 0 & 1 \end{bmatrix}.$$

The rotate transform is:

$$R = \begin{bmatrix} 1 & 0 & 0 & 0 \\ 0 & \cos 90° & -\sin 90° & 0 \\ 0 & \sin 90° & \cos 90° & 0 \\ 0 & 0 & 0 & 1 \end{bmatrix}, \quad \text{and} \quad R^{-1} = \begin{bmatrix} 1 & 0 & 0 & 0 \\ 0 & 0 & 1 & 0 \\ 0 & -1 & 0 & 0 \\ 0 & 0 & 0 & 1 \end{bmatrix}.$$

The translate transform is:

$$
\mathbf{T} = \begin{bmatrix} 1 & 0 & 0 & 0 \\ 0 & 1 & 0 & 0 \\ 0 & 0 & 1 & 10 \\ 0 & 0 & 0 & 1 \end{bmatrix}, \quad \text{and} \quad \mathbf{T}^{-1} = \begin{bmatrix} 1 & 0 & 0 & 0 \\ 0 & 1 & 0 & 0 \\ 0 & 0 & 1 & -10 \\ 0 & 0 & 0 & 1 \end{bmatrix}.
$$

The intersecting line is

$$
\mathbf{p} = \mathbf{t} + \lambda \mathbf{v}
$$

where,

$$
\mathbf{t} = 0\mathbf{i} + 0\mathbf{j} + 0\mathbf{k}, \quad \text{and} \quad \mathbf{v} = 0\mathbf{i} + 1\mathbf{j} + 1\mathbf{k}
$$

as shown in Fig. 10.17, from which we can predict that the line intersection points are $(0, 8, 8)$ and $(0, 12, 12)$.

Let's transform the line's vectors using:

$$
\mathbf{v}' = \mathbf{S}^{-1}\mathbf{R}^{-1}\mathbf{v}
$$
$$
\mathbf{t}' = \mathbf{S}^{-1}\mathbf{R}^{-1}\mathbf{T}^{-1}\mathbf{t}
$$
$$
\mathbf{p} = \mathbf{t}' + \lambda \mathbf{v}'. \tag{10.49}
$$

Note that $\mathbf{v}$ is not subjected to the translation transform. Therefore, (10.48) becomes

$$
\lambda = \frac{-(x_{v'}x_{t'} + y_{v'}y_{t'}) \pm \sqrt{r^2(x_{v'}^2 + y_{v'}^2) - \left| \begin{matrix} x_{v'} & y_{v'} \\ x_{t'} & y_{t'} \end{matrix} \right|^2}}{x_{v'}^2 + y_{v'}^2}. \tag{10.50}
$$

Calculate:

$$
\mathbf{S}^{-1}\mathbf{R}^{-1} = \begin{bmatrix} \frac{1}{2} & 0 & 0 & 0 \\ 0 & \frac{1}{2} & 0 & 0 \\ 0 & 0 & \frac{1}{2} & 0 \\ 0 & 0 & 0 & 1 \end{bmatrix} \begin{bmatrix} 1 & 0 & 0 & 0 \\ 0 & 0 & 1 & 0 \\ 0 & -1 & 0 & 0 \\ 0 & 0 & 0 & 1 \end{bmatrix} = \begin{bmatrix} \frac{1}{2} & 0 & 0 & 0 \\ 0 & 0 & \frac{1}{2} & 0 \\ 0 & -\frac{1}{2} & 0 & 0 \\ 0 & 0 & 0 & 1 \end{bmatrix}
$$

and

$$
\mathbf{S}^{-1}\mathbf{R}^{-1}\mathbf{T}^{-1} = \begin{bmatrix} \frac{1}{2} & 0 & 0 & 0 \\ 0 & 0 & \frac{1}{2} & 0 \\ 0 & -\frac{1}{2} & 0 & 0 \\ 0 & 0 & 0 & 1 \end{bmatrix} \begin{bmatrix} 1 & 0 & 0 & 0 \\ 0 & 1 & 0 & 0 \\ 0 & 0 & 1 & -10 \\ 0 & 0 & 0 & 1 \end{bmatrix} = \begin{bmatrix} \frac{1}{2} & 0 & 0 & 0 \\ 0 & 0 & \frac{1}{2} & -5 \\ 0 & -\frac{1}{2} & 0 & 0 \\ 0 & 0 & 0 & 1 \end{bmatrix}.
$$

Therefore,

$$\mathbf{t'} = \begin{bmatrix} \frac{1}{2} & 0 & 0 & 0 \\ 0 & 0 & \frac{1}{2} & -5 \\ 0 & -\frac{1}{2} & 0 & 0 \\ 0 & 0 & 0 & 1 \end{bmatrix} \begin{bmatrix} 0 \\ 0 \\ 0 \\ 1 \end{bmatrix} = \begin{bmatrix} 0 \\ -5 \\ 0 \\ 1 \end{bmatrix}$$

and

$$\mathbf{v'} = \begin{bmatrix} \frac{1}{2} & 0 & 0 & 0 \\ 0 & 0 & \frac{1}{2} & 0 \\ 0 & -\frac{1}{2} & 0 & 0 \\ 0 & 0 & 0 & 1 \end{bmatrix} \begin{bmatrix} 0 \\ 1 \\ 1 \\ 1 \end{bmatrix} = \begin{bmatrix} 0 \\ \frac{1}{2} \\ -\frac{1}{2} \\ 1 \end{bmatrix}.$$

Substituting $\mathbf{t'}$ and $\mathbf{v'}$ in (10.50) we get:

$$\lambda = \frac{-(0 - 2\frac{1}{2}) \pm \sqrt{\frac{1}{4} - \begin{vmatrix} 0 & \frac{1}{2} \\ 0 & -5 \end{vmatrix}^2}}{\frac{1}{4}}$$

$$= \frac{2\frac{1}{2} \pm \frac{1}{2}}{\frac{1}{4}} = 8, \quad \text{and} \quad 12.$$

Substituting λ in (10.49) we have:

$$\mathbf{p} = \mathbf{0} + 8(0\mathbf{i} + 1\mathbf{j} + 1\mathbf{k})$$
$$(x_p, y_p, z_p) = (0, 8, 8)$$
$$\mathbf{p} = \mathbf{0} + 12(0\mathbf{i} + 1\mathbf{j} + 1\mathbf{k})$$
$$(x_p, y_p, z_p) = (0, 12, 12).$$

As predicted!

What has been covered so far assumes that the cylinder has an infinite extent along the z-axis, which is not always the case. Normally, cylinders have a finite length, and a mechanism is required to set these physical limits. One convenient method is to identify the minimum and maximum z-extents of the cylinder, z_{min} and z_{max}, respectively. And when we calculate the two possible intersection values for λ, e.g. λ_1 and λ_2, we can test whether the z-component of $\mathbf{p}$ is within the defined range:

$$z_1 = z_t + \lambda_1 z_v$$
$$z_2 = z_t + \lambda_2 z_v$$

by making sure that:

$$z_{min} < (z_1, z_2) < z_{max}.$$

Fig. 10.18 A cylinder's end
caps

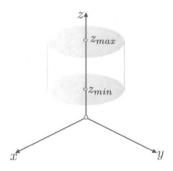

Another assumption with the above analysis is that the cylinder is open either end. If end caps are required, then plane equations can be placed at z_{min} and z_{max}, as shown in Fig. 10.18.

We also need to recognise the conditions when such an intersection can arise. An intersection with the z_{min} end cap occurs when z_1 and z_2 are either side of the z_{min} plane. Similarly, an intersection with the z_{max} end cap occurs when z_1 and z_2 are either side of the z_{max} plane.

We have already examined the intersection of a line with a plane, and discovered that given a line:

$$\mathbf{p} = \mathbf{t} + \lambda \mathbf{v}$$

and a plane equation:

$$ax + by + cz = d$$

where,

$$\mathbf{n} = a\mathbf{i} + b\mathbf{j} + c\mathbf{k}$$

that,

$$\lambda = \frac{-(\mathbf{n} \cdot \mathbf{t} + d)}{\mathbf{n} \cdot \mathbf{v}}.$$

If we place a plane at z_{min}, we have

$$d = z_{min}, \quad \text{and} \quad \mathbf{n} = -\mathbf{k}$$

and

$$\lambda_3 = \frac{-(-z_t + z_{min})}{-z_v} = \frac{z_{min} - z_t}{z_v}.$$

λ_3 can now be used to reveal the intersection point using

$$\mathbf{p} = \mathbf{t} + \lambda_3 \mathbf{v}.$$

Similarly, if we place a plane at z_{max}, we have

$$\lambda_4 = \frac{z_{\text{max}} - z_t}{z_v}$$

and

$$\mathbf{p} = \mathbf{t} + \lambda_4 \mathbf{v}.$$

Note that both end cap normal vectors point in the same direction.

10.10 A Line Intersecting a Cone

Another geometric primitive is the cone, and once again this is normally aligned, like the cylinder, with the z-axis, as shown in Fig. 10.19. A simple form of the cone equation is

$$z^2 = x^2 + y^2$$

which makes the internal angle of the apex $90°$.

The line equation is given by

$$\mathbf{p} = \mathbf{t} + \lambda \mathbf{v}.$$

Then for an intersection at P, we have:

$$z_p^2 = x_p^2 + y_p^2$$

which means that

$$(z_t + \lambda z_v)^2 = (x_t + \lambda x_v)^2 + (y_t + \lambda y_v)^2. \tag{10.51}$$

Fig. 10.19 A line
intersecting a cone

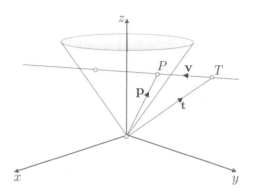

Expanding and simplifying (10.51):

$$z_t^2 + 2\lambda z_t z_v + \lambda^2 z_v^2 = x_t^2 + 2\lambda x_t x_v + \lambda^2 x_v^2 + y_t^2 + 2\lambda y_t y_v + \lambda^2 y_v^2$$
$$0 = \lambda^2 (x_v^2 + y_v^2 - z_v^2) + \lambda(2x_t x_v + 2y_t y_v - 2z_t z_v)$$
$$+ x_t^2 + y_t^2 - z_t^2$$

which is a quadratic in λ, and solved using

$$\lambda = \frac{-B \pm \sqrt{B^2 - 4AC}}{2A}$$

where,

$$A = x_v^2 + y_v^2 - z_v^2, \quad B = 2x_t x_v + 2y_t y_v - 2z_t z_v, \quad C = x_t^2 + y_t^2 - z_t^2.$$

Thus:

$$\lambda = \frac{-(x_t x_v + y_t y_v - z_t z_v) \pm \sqrt{(x_t x_v + y_t y_v - z_t z_v)^2 - (x_v^2 + y_v^2 - z_v^2)\left(x_t^2 + y_t^2 - z_t^2\right)}}{x_v^2 + y_v^2 - z_v^2}$$

which reduces to

$$\lambda = \frac{-(x_t x_v + y_t y_v - z_t z_v) \pm \sqrt{\begin{vmatrix} y_v & z_v \\ y_t & z_t \end{vmatrix}^2 + \begin{vmatrix} z_v & x_v \\ z_t & x_t \end{vmatrix}^2 - \begin{vmatrix} x_v & y_v \\ x_t & y_t \end{vmatrix}^2}}{x_v^2 + y_v^2 - z_v^2}. \qquad (10.52)$$

The value of the discriminant of (10.52) determines whether the line misses, touches or intersects the cone:

Miss condition:

$$\begin{vmatrix} y_v & z_v \\ y_t & z_t \end{vmatrix}^2 + \begin{vmatrix} z_v & x_v \\ z_t & x_t \end{vmatrix}^2 - \begin{vmatrix} x_v & y_v \\ x_t & y_t \end{vmatrix}^2 < 0.$$

Touch condition:

$$\begin{vmatrix} y_v & z_v \\ y_t & z_t \end{vmatrix}^2 + \begin{vmatrix} z_v & x_v \\ z_t & x_t \end{vmatrix}^2 - \begin{vmatrix} x_v & y_v \\ x_t & y_t \end{vmatrix}^2 = 0.$$

Intersect condition:

$$\begin{vmatrix} y_v & z_v \\ y_t & z_t \end{vmatrix}^2 + \begin{vmatrix} z_v & x_v \\ z_t & x_t \end{vmatrix}^2 - \begin{vmatrix} x_v & y_v \\ x_t & y_t \end{vmatrix}^2 > 0.$$

If the denominator of (10.52) is zero, the ray is parallel with the cone's side.

Limits to the cone's length can be set using the same technique for the cylinder, where z_{min} and z_{max} define the z-extent of the cone. These can be negative or positive, and determine whether the cone is aligned with the negative or positive z-axis, respectively.

If the line intersects the cone twice, there will be two values of λ: λ_1 and λ_2, which generate the two z coordinates of the intersections:

$$z_1 = z_t + \lambda_1 z_v$$
$$z_2 = z_t + \lambda_2 z_v.$$

Once again, to ensure that the intersections occur in the isolated part of the cone, we test for

$$z_{min} < (z_1, \ z_2) < z_{max}.$$

By making either z_{min} or z_{max} equal zero retains the cone's apex. However, if they are both negative or positive, a truncated cone is created. And if the cone requires end caps, the same technique used for capping cylinders can be applied to cones.

10.11   A Line Intersecting a Triangle

Triangles play an important role in modeling CG objects, and the line-triangle intersection problem is important in the design of renderers. Triangles are planar and it is easy to compute whether, and where, a line intersects the plane shared by the triangle. What is not so easy is proving that the intersection is within the triangle's boundary. Fortunately for us, in 1827, August Ferdinand Möbius [1790–1868] published *Der barycentrische Calcul*, which formalised the analysis of centers of gravity, and also laid the foundations for vectors. Today, the work of Möbius is called *barycentric coordinates*.

Barycentric coordinates provide the key to unlocking this problem, and if you have never used them, refer to the author's book [1].

We begin by defining triangle $\Delta P_1 P_2 P_3$ with vertices P_1, P_2 and P_3 and their corresponding position vectors $\mathbf{p}_1$, $\mathbf{p}_2$ and $\mathbf{p}_3$. This scenario is shown in Fig. 10.20, with the following points and vectors:

- $P_1(x_1, y_1, z_1)$ is a vertex of the triangle,
- $\mathbf{p}_1 = x_1\mathbf{i} + y_1\mathbf{j} + z_1\mathbf{k}$ is P_1's position vector,
- $P_2(x_2, y_2, z_2)$ is a vertex of the triangle,
- $\mathbf{p}_2 = x_2\mathbf{i} + y_2\mathbf{j} + z_2\mathbf{k}$ is P_2's position vector,
- $P_3(x_3, y_3, z_3)$ is a vertex of the triangle,
- $\mathbf{p}_3 = x_3\mathbf{i} + y_3\mathbf{j} + z_3\mathbf{k}$ is P_3's position vector,
- $P(x, y, z)$ is a point on the triangle,
- $\mathbf{p} = x\mathbf{i} + y\mathbf{j} + z\mathbf{k}$ is P's position vector,
- $T(x_t, y_t, z_t)$ is a point on the line,

Fig. 10.20 A line
intersecting a triangle

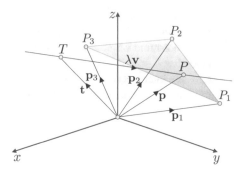

• $\mathbf{t} = x_t\mathbf{i} + y_t\mathbf{j} + z_t\mathbf{k}$ is T's position vector,
• $\mathbf{v} = x_v\mathbf{i} + y_v\mathbf{j} + z_v\mathbf{k}$ is the line's direction vector,
• $\mathbf{p} = \mathbf{t} + \lambda\mathbf{v}$ is the equation of the line.

We can state that for any point $P(x_p, y_p, z_p)$ on triangle $\Delta P_1 P_2 P_3$ is given by

$$\mathbf{p}(r, s) = q\mathbf{p}_1 + r\mathbf{p}_2 + s\mathbf{p}_3$$

where $\mathbf{p}$ is P's position vector. The scalars q, r and s are barycentric coordinates and satisfy the following conditions:

$$q + r + s = 1, \quad (q, r, s) \in [0, 1]$$

which makes

$$q = 1 - r - s.$$

If the intersecting line is given by:

$$\mathbf{p} = \mathbf{t} + \lambda\mathbf{v}.$$

Then,

$$
\begin{aligned}
\mathbf{t} + \lambda\mathbf{v} &= (1 - r - s)\mathbf{p}_1 + r\mathbf{p}_2 + s\mathbf{p}_3 \\
&= \mathbf{p}_1 - r\mathbf{p}_1 - s\mathbf{p}_1 + r\mathbf{p}_2 + s\mathbf{p}_3 \\
&= \mathbf{p}_1 + r(\mathbf{p}_2 - \mathbf{p}_1) + s(\mathbf{p}_3 - \mathbf{p}_1) \\
\mathbf{t} - \mathbf{p}_1 &= -\lambda\mathbf{v} + r(\mathbf{p}_2 - \mathbf{p}_1) + s(\mathbf{p}_3 - \mathbf{p}_1)
\end{aligned}
$$

and

$$\mathbf{t} - \mathbf{p}_1 = \begin{bmatrix} -\mathbf{v} & \mathbf{p}_2 - \mathbf{p}_1 & \mathbf{p}_3 - \mathbf{p}_1 \end{bmatrix} \begin{bmatrix} \lambda \\ r \\ s \end{bmatrix}. \tag{10.53}$$

Solving (10.53) reveals the value of λ, r and s. The barycentric coordinates r and s tell us whether the point is on the triangle, and λ is the position along $\mathbf{v}$. First, though, we need to solve (10.53).

As an aside, consider the simultaneous equations:

$$c_1 = a_1x + b_1y \tag{10.54}$$
$$c_2 = a_2x + b_2y. \tag{10.55}$$

We can remove y from (10.54) and (10.55) by multiplying (10.54) by b_2, (10.55) by $-b_1$, and adding:

$$b_2c_1 = a_1b_2x + b_1b_2y$$
$$-b_1c_2 = -a_2b_1x - b_1b_2y$$
$$b_2c_1 - b_1c_2 = a_1b_2x - a_2b_1x$$

$$x = \frac{b_2c_1 - b_1c_2}{a_1b_2 - a_2b_1} = \frac{\begin{vmatrix} c_1 & b_1 \\ c_2 & b_2 \end{vmatrix}}{\begin{vmatrix} a_1 & b_1 \\ a_2 & b_2 \end{vmatrix}}.$$

Similarly, we can remove x from (10.54) and (10.55) by multiplying (10.54) by $-a_2$, (10.55) by a_1, and adding:

$$-a_2c_1 = -a_1a_2x - a_2b_1y$$
$$a_1c_2 = a_1a_2x + a_1b_2y$$
$$a_1c_2 - a_2c_1 = a_1b_2y - a_2b_1y$$

$$y = \frac{a_1c_2 - a_2c_1}{a_1b_2 - a_2b_1} = \frac{\begin{vmatrix} a_1 & c_1 \\ a_2 & c_2 \end{vmatrix}}{\begin{vmatrix} a_1 & b_1 \\ a_2 & b_2 \end{vmatrix}}.$$

This symmetry was discovered by the Swiss mathematician Gabriel Cramer [1704–1752] and is now known as *Cramer's Rule*; however, it only works when the determinant in the denominator is non zero.

Similarly, for three simultaneous equations:

$$d_1 = a_1x + b_1y + c_1z$$
$$d_2 = a_2x + b_2y + c_2z$$
$$d_3 = a_3x + b_3y + c_3z.$$

Therefore,

$$x = \frac{\begin{vmatrix} d_1 & b_1 & c_1 \\ d_2 & b_2 & c_2 \\ d_3 & b_3 & c_3 \end{vmatrix}}{\begin{vmatrix} a_1 & b_1 & c_1 \\ a_2 & b_2 & c_2 \\ a_3 & b_3 & c_3 \end{vmatrix}}, \quad y = \frac{\begin{vmatrix} a_1 & d_1 & c_1 \\ a_2 & d_2 & c_2 \\ a_3 & d_3 & c_3 \end{vmatrix}}{\begin{vmatrix} a_1 & b_1 & c_1 \\ a_2 & b_2 & c_2 \\ a_3 & b_3 & c_3 \end{vmatrix}}, \quad z = \frac{\begin{vmatrix} a_1 & b_1 & d_1 \\ a_2 & b_2 & d_2 \\ a_3 & b_3 & d_3 \end{vmatrix}}{\begin{vmatrix} a_1 & b_1 & c_1 \\ a_2 & b_2 & c_2 \\ a_3 & b_3 & c_3 \end{vmatrix}}.$$

Before applying Cramer's Rule, let's tidy up (10.53) as follows:

$$\mathbf{k} = \mathbf{t} - \mathbf{p}_1, \qquad \mathbf{p}_{21} = \mathbf{p}_2 - \mathbf{p}_1, \qquad \mathbf{p}_{31} = \mathbf{p}_3 - \mathbf{p}_1.$$

Therefore,

$$\mathbf{k} = \begin{bmatrix} -\mathbf{v} & \mathbf{p}_{23} & \mathbf{p}_{31} \end{bmatrix} \begin{bmatrix} \lambda \\ r \\ s \end{bmatrix} \tag{10.56}$$

and

$$\lambda = \frac{\begin{vmatrix} x_k & x_{21} & x_{31} \\ y_k & y_{21} & y_{31} \\ z_k & z_{21} & z_{31} \end{vmatrix}}{DET}, \quad r = \frac{\begin{vmatrix} -x_v & x_k & x_{31} \\ -y_v & y_k & y_{31} \\ -z_v & z_k & z_{31} \end{vmatrix}}{DET}, \quad s = \frac{\begin{vmatrix} -x_v & x_{21} & x_k \\ -y_v & y_{21} & y_k \\ -z_v & z_{21} & z_k \end{vmatrix}}{DET} \tag{10.57}$$

where,

$$DET = \begin{vmatrix} -x_v & x_{21} & x_{31} \\ -y_v & y_{21} & y_{31} \\ -z_v & z_{21} & z_{31} \end{vmatrix}.$$

There is an alternative solution using the scalar triple product:

$$\begin{vmatrix} x_a & y_a & z_a \\ x_b & y_b & z_b \\ x_c & y_c & z_c \end{vmatrix} = \mathbf{a} \cdot (\mathbf{b} \times \mathbf{c}) = \mathbf{b} \cdot (\mathbf{c} \times \mathbf{a}) = \mathbf{c} \cdot (\mathbf{a} \times \mathbf{b})$$

$$= -\mathbf{a} \cdot (\mathbf{c} \times \mathbf{b}) = -\mathbf{b} \cdot (\mathbf{a} \times \mathbf{c}) = -\mathbf{c} \cdot (\mathbf{b} \times \mathbf{a}).$$

Applying this relationships to (10.56), we obtain:

$$\lambda = \frac{\begin{vmatrix} \mathbf{k} & \mathbf{p}_{21} & \mathbf{p}_{31} \end{vmatrix}}{\begin{vmatrix} -\mathbf{v} & \mathbf{p}_{21} & \mathbf{p}_{31} \end{vmatrix}} = \frac{(\mathbf{k} \times \mathbf{p}_{21}) \cdot \mathbf{p}_{31}}{(\mathbf{v} \times \mathbf{p}_{31}) \cdot \mathbf{p}_{21}} \tag{10.58}$$

$$r = \frac{\begin{vmatrix} -\mathbf{v} & \mathbf{k} & \mathbf{p}_{31} \end{vmatrix}}{\begin{vmatrix} -\mathbf{v} & \mathbf{p}_{21} & \mathbf{p}_{31} \end{vmatrix}} = \frac{(\mathbf{v} \times \mathbf{p}_{31}) \cdot \mathbf{k}}{(\mathbf{v} \times \mathbf{p}_{31}) \cdot \mathbf{p}_{21}} \tag{10.59}$$

$$s = \frac{\begin{vmatrix} -\mathbf{v} & \mathbf{p}_{21} & \mathbf{k} \end{vmatrix}}{\begin{vmatrix} -\mathbf{v} & \mathbf{p}_{21} & \mathbf{p}_{31} \end{vmatrix}} = \frac{(\mathbf{k} \times \mathbf{p}_{21}) \cdot \mathbf{v}}{(\mathbf{v} \times \mathbf{p}_{31}) \cdot \mathbf{p}_{21}}. \tag{10.60}$$

Fig. 10.21 A line
intersecting a triangle at P_2

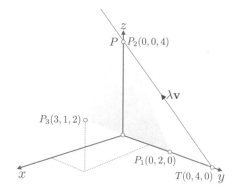

Note that the equations (10.58), (10.59), (10.60), contain two common terms: $\mathbf{k} \times \mathbf{p}_{21}$
and $\mathbf{v} \times \mathbf{p}_{31}$. If these are computed separately, we can write:

$$\lambda = \frac{\mathbf{m} \cdot \mathbf{p}_{31}}{\mathbf{n} \cdot \mathbf{p}_{21}}, \quad r = \frac{\mathbf{n} \cdot \mathbf{k}}{\mathbf{n} \cdot \mathbf{p}_{21}}, \quad s = \frac{\mathbf{m} \cdot \mathbf{v}}{\mathbf{n} \cdot \mathbf{p}_{21}}$$

where,

$$\mathbf{m} = \mathbf{k} \times \mathbf{p}_{21}, \quad \mathbf{n} = \mathbf{v} \times \mathbf{p}_{31}.$$

Let's put the above theory into practice using a triangle formed from the points
$P_1(0, 2, 0)$, $P_2(0, 0, 4)$ and $P_3(3, 1, 2)$, with the line $\mathbf{p} = \mathbf{t} + \lambda \mathbf{v}$, where

$$\mathbf{t} = 0\mathbf{i} + 4\mathbf{j} + 0\mathbf{k}, \quad \mathbf{v} = 0\mathbf{i} - 1\mathbf{j} + 1\mathbf{k}$$

and intersects the triangle at P_2 as shown in Fig. 10.21.
Therefore,

$$\mathbf{k} = \mathbf{t} - \mathbf{p}_1 = (0\mathbf{i} + 4\mathbf{j} + 0\mathbf{k}) - (0\mathbf{i} + 2\mathbf{j} + 0\mathbf{k}) = 0\mathbf{i} + 2\mathbf{j} + 0\mathbf{k},$$
$$\mathbf{p}_{21} = \mathbf{p}_2 - \mathbf{p}_1 = (0\mathbf{i} + 0\mathbf{j} + 4\mathbf{k}) - (0\mathbf{i} + 2\mathbf{j} + 0\mathbf{k}) = 0\mathbf{i} - 2\mathbf{j} + 4\mathbf{k}$$
$$\mathbf{p}_{31} = \mathbf{p}_3 - \mathbf{p}_1 = (3\mathbf{i} + 1\mathbf{j} + 2\mathbf{k}) - (0\mathbf{i} + 2\mathbf{j} + 0\mathbf{k}) = 3\mathbf{i} - 1\mathbf{j} + 2\mathbf{k}.$$

Using (10.57) we obtain:

$$DET = \begin{vmatrix} -x_v & x_{21} & x_{31} \\ -y_v & y_{21} & y_{31} \\ -z_v & z_{21} & z_{31} \end{vmatrix} = \begin{vmatrix} 0 & 0 & 3 \\ 1 & -2 & -1 \\ -1 & 4 & 2 \end{vmatrix} = 6$$

$$\lambda = \frac{\begin{vmatrix} x_k & x_{21} & x_{31} \\ y_k & y_{21} & y_{31} \\ z_k & z_{21} & z_{31} \end{vmatrix}}{DET} = \frac{\begin{vmatrix} 0 & 0 & 3 \\ 2 & -2 & -1 \\ 0 & 4 & 2 \end{vmatrix}}{6} = 4$$

$$r = \frac{\begin{vmatrix} -x_v & x_k & x_{31} \\ -y_v & y_k & y_{31} \\ -z_v & z_k & z_{31} \end{vmatrix}}{DET} = \frac{\begin{vmatrix} 0 & 0 & 3 \\ 1 & 2 & -1 \\ -1 & 0 & 2 \end{vmatrix}}{6} = 1$$

$$s = \frac{\begin{vmatrix} -x_v & x_{21} & x_k \\ -y_v & y_{21} & y_k \\ -z_v & z_{21} & z_k \end{vmatrix}}{DET} = \frac{\begin{vmatrix} 0 & 0 & 0 \\ 1 & -2 & 2 \\ -1 & 4 & 0 \end{vmatrix}}{6} = 0.$$

$r = 1$ confirms that the line intersects the triangle at P_2. Similarly, with $\lambda = 4$, the line intersects the point:

$$\mathbf{p} = (0\mathbf{i} + 4\mathbf{j} + 0\mathbf{k}) + 4(0\mathbf{i} - 1\mathbf{j} + 1\mathbf{k}) = 0\mathbf{i} + 0\mathbf{j} + 4\mathbf{k}$$

which is P_2.

Using (10.58), (10.59) and (10.60) we have:

$$\mathbf{m} = \begin{vmatrix} \mathbf{i} & \mathbf{j} & \mathbf{k} \\ x_k & y_k & z_k \\ x_{21} & y_{21} & z_{21} \end{vmatrix} = \begin{vmatrix} \mathbf{i} & \mathbf{j} & \mathbf{k} \\ 0 & 2 & 0 \\ 0 & -2 & 4 \end{vmatrix} = 8\mathbf{i} + 0\mathbf{j} + 0\mathbf{k}$$

$$\mathbf{n} = \begin{vmatrix} \mathbf{i} & \mathbf{j} & \mathbf{k} \\ x_v & y_v & z_v \\ x_{31} & y_{31} & z_{31} \end{vmatrix} = \begin{vmatrix} \mathbf{i} & \mathbf{j} & \mathbf{k} \\ 0 & -1 & 1 \\ 3 & -1 & 2 \end{vmatrix} = -1\mathbf{i} + 3\mathbf{j} + 3\mathbf{k}$$

$$\mathbf{n} \cdot \mathbf{p}_{21} = (-1\mathbf{i} + 3\mathbf{j} + 3\mathbf{k}) \cdot (0\mathbf{i} - 2\mathbf{j} + 4\mathbf{k}) = 6$$

$$\lambda = \frac{\mathbf{m} \cdot \mathbf{p}_{31}}{\mathbf{n} \cdot \mathbf{p}_{21}} = \frac{(8\mathbf{i} + 0\mathbf{j} + 0\mathbf{k}) \cdot (3\mathbf{i} - 1\mathbf{j} + 2\mathbf{k})}{6} = 4$$

$$r = \frac{\mathbf{n} \cdot \mathbf{k}}{\mathbf{n} \cdot \mathbf{p}_{21}} = \frac{(-1\mathbf{i} + 3\mathbf{j} + 3\mathbf{k}) \cdot (0\mathbf{i} + 2\mathbf{j} + 0\mathbf{k})}{6} = 1$$

$$s = \frac{\mathbf{m} \cdot \mathbf{v}}{\mathbf{n} \cdot \mathbf{p}_{21}} = \frac{(8\mathbf{i} + 0\mathbf{j} + 0\mathbf{k}) \cdot (0\mathbf{i} - 1\mathbf{j} + 1\mathbf{k})}{6} = 0$$

which are identical to the previous results. Now let's try another line:

$$\mathbf{t} = 0\mathbf{i} + 4\mathbf{j} + 0\mathbf{k}, \quad \mathbf{v} = 1\mathbf{i} - 1\mathbf{j} + 1\mathbf{k}.$$

Using (10.57), we obtain:

$$\mathbf{k} = \mathbf{t} - \mathbf{p}_1 = (0\mathbf{i} + 4\mathbf{j} + 0\mathbf{k}) - (0\mathbf{i} + 2\mathbf{j} + 0\mathbf{k}) = 0\mathbf{i} + 2\mathbf{j} + 0\mathbf{k}$$

$$DET = \begin{vmatrix} -x_v & x_{21} & x_{31} \\ -y_v & y_{21} & y_{31} \\ -z_v & z_{21} & z_{31} \end{vmatrix} = \begin{vmatrix} 1 & 0 & 3 \\ -1 & -2 & -1 \\ 1 & 4 & 2 \end{vmatrix} = 6$$

$$\lambda = \frac{\begin{vmatrix} x_k & x_{21} & x_{31} \\ y_k & y_{21} & y_{31} \\ z_k & z_{21} & z_{31} \end{vmatrix}}{DET} = \frac{\begin{vmatrix} 0 & 0 & 3 \\ 2 & -2 & -1 \\ 0 & 4 & 2 \end{vmatrix}}{6} = 4$$

$$r = \frac{\begin{vmatrix} -x_v & x_k & x_{31} \\ -y_v & y_k & y_{31} \\ -z_v & z_k & z_{31} \end{vmatrix}}{DET} = \frac{\begin{vmatrix} 1 & 0 & 3 \\ -1 & 2 & -1 \\ 1 & 0 & 2 \end{vmatrix}}{6} = \frac{1}{3}$$

$$s = \frac{\begin{vmatrix} -x_v & x_{21} & x_k \\ -y_v & y_{21} & y_k \\ -z_v & z_{21} & z_k \end{vmatrix}}{DET} = \frac{\begin{vmatrix} -1 & 0 & 0 \\ 1 & -2 & 2 \\ -1 & 4 & 0 \end{vmatrix}}{6} = 1\frac{1}{3}.$$

As the value of $s > 1$, the intersection point is outside the triangle.

Let's make sure that we obtain the same result using (10.58), (10.59) and (10.60):

$$\mathbf{m} = \begin{vmatrix} \mathbf{i} & \mathbf{j} & \mathbf{k} \\ x_k & y_k & z_k \\ x_{21} & y_{21} & z_{21} \end{vmatrix} = \begin{vmatrix} \mathbf{i} & \mathbf{j} & \mathbf{k} \\ 0 & 2 & 0 \\ 0 & -2 & 4 \end{vmatrix} = 8\mathbf{i} + 0\mathbf{j} + 0\mathbf{k}$$

$$\mathbf{n} = \begin{vmatrix} \mathbf{i} & \mathbf{j} & \mathbf{k} \\ x_v & y_v & z_v \\ x_{31} & y_{31} & z_{31} \end{vmatrix} = \begin{vmatrix} \mathbf{i} & \mathbf{j} & \mathbf{k} \\ 1 & -1 & 1 \\ 3 & -1 & 2 \end{vmatrix} = -1\mathbf{i} + 1\mathbf{j} + 2\mathbf{k}$$

$$\mathbf{n} \cdot \mathbf{p}_{21} = (-1\mathbf{i} + 1\mathbf{j} + 2\mathbf{k}) \cdot (0\mathbf{i} - 2\mathbf{j} + 4\mathbf{k}) = 6$$

$$\lambda = \frac{\mathbf{m} \cdot \mathbf{p}_{31}}{\mathbf{n} \cdot \mathbf{p}_{21}} = \frac{(8\mathbf{i} + 0\mathbf{j} + 0\mathbf{k}) \cdot (3\mathbf{i} - 1\mathbf{j} + 2\mathbf{k})}{6} = 4$$

$$r = \frac{\mathbf{n} \cdot \mathbf{k}}{\mathbf{n} \cdot \mathbf{p}_{21}} = \frac{(-1\mathbf{i} + 1\mathbf{j} + 2\mathbf{k}) \cdot (0\mathbf{i} + 2\mathbf{j} + 0\mathbf{k})}{6} = \frac{1}{3}$$

$$s = \frac{\mathbf{m} \cdot \mathbf{v}}{\mathbf{n} \cdot \mathbf{p}_{21}} = \frac{(8\mathbf{i} + 0\mathbf{j} + 0\mathbf{k}) \cdot (1\mathbf{i} - 1\mathbf{j} + 1\mathbf{k})}{6} = 1\frac{1}{3}$$

which confirm the previous values.

Finally, let's select a line that intersects the triangle:

$$\mathbf{t} = 0\mathbf{i} + 4\mathbf{j} + 0\mathbf{k}, \quad \mathbf{v} = 2\mathbf{i} - 4\mathbf{j} + 2\mathbf{k}.$$

Using (10.57), we obtain:

$$\mathbf{k} = \mathbf{t} - \mathbf{p}_1 = (0\mathbf{i} + 4\mathbf{j} + 0\mathbf{k}) - (0\mathbf{i} + 2\mathbf{j} + 0\mathbf{k}) = 0\mathbf{i} + 2\mathbf{j} + 0\mathbf{k}$$

$$DET = \begin{vmatrix} -x_v & x_{21} & x_{31} \\ -y_v & y_{21} & y_{31} \\ -z_v & z_{21} & z_{31} \end{vmatrix} = \begin{vmatrix} -2 & 0 & 3 \\ 4 & -2 & -1 \\ -2 & 4 & 2 \end{vmatrix} = 36$$

$$\lambda = \frac{\begin{vmatrix} x_k & x_{21} & x_{31} \\ y_k & y_{21} & y_{31} \\ z_k & z_{21} & z_{31} \end{vmatrix}}{DET} = \frac{\begin{vmatrix} 0 & 0 & 3 \\ 2 & -2 & -1 \\ 0 & 4 & 2 \end{vmatrix}}{36} = \frac{2}{3}$$

$$r = \frac{\begin{vmatrix} -x_v & x_k & x_{31} \\ -y_v & y_k & y_{31} \\ -z_v & z_k & z_{31} \end{vmatrix}}{DET} = \frac{\begin{vmatrix} -2 & 0 & 3 \\ 4 & 2 & -1 \\ -2 & 0 & 2 \end{vmatrix}}{36} = \frac{1}{9}$$

$$s = \frac{\begin{vmatrix} -x_v & x_{21} & x_k \\ -y_v & y_{21} & y_k \\ -z_v & z_{21} & z_k \end{vmatrix}}{DET} = \frac{\begin{vmatrix} -2 & 0 & 0 \\ 4 & -2 & 2 \\ -2 & 4 & 0 \end{vmatrix}}{36} = \frac{4}{9}.$$

$r = \frac{1}{9}$, and $s = \frac{4}{9}$, confirms that the line intersects the triangle. The point of intersection is given by:

$$\mathbf{p} = \mathbf{t} + \lambda \mathbf{v}$$
$$= (0\mathbf{i} + 4\mathbf{j} + 0\mathbf{k}) + \tfrac{2}{3}(2\mathbf{i} - 4\mathbf{j} + 2\mathbf{k})$$
$$= 4\mathbf{j} + \left(\tfrac{4}{3}\mathbf{i} - \tfrac{8}{3}\mathbf{j} + \tfrac{4}{3}\mathbf{k} \right)$$
$$P = \left(\tfrac{4}{3}, \tfrac{4}{3}, \tfrac{4}{3} \right).$$

Using (10.58), (10.59) and (10.60):

$$\mathbf{k} = \mathbf{t} - \mathbf{p}_1 = (0\mathbf{i} + 4\mathbf{j} + 0\mathbf{k}) - (0\mathbf{i} + 2\mathbf{j} + 0\mathbf{k}) = 0\mathbf{i} + 2\mathbf{j} + 0\mathbf{k}$$

$$\mathbf{m} = \begin{vmatrix} \mathbf{i} & \mathbf{j} & \mathbf{k} \\ x_k & y_k & z_k \\ x_{21} & y_{21} & z_{21} \end{vmatrix} = \begin{vmatrix} \mathbf{i} & \mathbf{j} & \mathbf{k} \\ 0 & 2 & 0 \\ 0 & -2 & 4 \end{vmatrix} = 8\mathbf{i} + 0\mathbf{j} + 0\mathbf{k}$$

$$\mathbf{n} = \begin{vmatrix} \mathbf{i} & \mathbf{j} & \mathbf{k} \\ x_v & y_v & z_v \\ x_{31} & y_{31} & z_{31} \end{vmatrix} = \begin{vmatrix} \mathbf{i} & \mathbf{j} & \mathbf{k} \\ 2 & -4 & 2 \\ 3 & -1 & 2 \end{vmatrix} = -6\mathbf{i} + 2\mathbf{j} + 10\mathbf{k}$$

$$\mathbf{n} \cdot \mathbf{p}_{21} = (-6\mathbf{i} + 2\mathbf{j} + 10\mathbf{k}) \cdot (0\mathbf{i} - 2\mathbf{j} + 4\mathbf{k}) = 36$$

$$\lambda = \frac{\mathbf{m} \cdot \mathbf{p}_{31}}{\mathbf{n} \cdot \mathbf{p}_{21}} = \frac{(8\mathbf{i} + 0\mathbf{j} + 0\mathbf{k}) \cdot (3\mathbf{i} - 1\mathbf{j} + 2\mathbf{k})}{36} = \frac{2}{3}$$

$$r = \frac{\mathbf{n} \cdot \mathbf{k}}{\mathbf{n} \cdot \mathbf{p}_{21}} = \frac{(-6\mathbf{i} + 2\mathbf{j} + 10\mathbf{k}) \cdot (0\mathbf{i} + 2\mathbf{j} + 0\mathbf{k})}{36} = \frac{1}{9}$$

$$s = \frac{\mathbf{m} \cdot \mathbf{v}}{\mathbf{n} \cdot \mathbf{p}_{21}} = \frac{(8\mathbf{i} + 0\mathbf{j} + 0\mathbf{k}) \cdot (2\mathbf{i} - 4\mathbf{j} + 2\mathbf{k})}{36} = \frac{4}{9}$$

which are identical to the previous results.

10.12   A Point Inside a Triangle

Having discovered how to test whether a line intersects a triangle, it is an opportune moment to describe how to test whether a point is inside or outside a triangle. Once again, we draw upon barycentric coordinates to reveal the solution. Figure 10.22 shows a triangle $\Delta P_1 P_2 P_3$ with the following points and vectors:

- $P_1(x_1, y_1, z_1)$ is a vertex of the triangle,
- $\mathbf{p}_1 = x_1\mathbf{i} + y_1\mathbf{j} + z_1\mathbf{k}$ is P_1's position vector,
- $P_2(x_2, y_2, z_2)$ is a vertex of the triangle,
- $\mathbf{p}_2 = x_2\mathbf{i} + y_2\mathbf{j} + z_2\mathbf{k}$ is P_2's position vector,
- $P_3(x_3, y_3, z_3)$ is a vertex of the triangle,
- $\mathbf{p}_3 = x_3\mathbf{i} + y_3\mathbf{j} + z_3\mathbf{k}$ is P_3's position vector,
- $P(x, y, z)$ is a point on the triangle,
- $\mathbf{p} = x\mathbf{i} + y\mathbf{j} + z\mathbf{k}$ is P's position vector.

Given a point P, which resides on the plane containing $\Delta P_1 P_2 P_3$, we can state that,

$$\mathbf{p} = r\mathbf{p}_1 + s\mathbf{p}_2 + t\mathbf{p}_3 \tag{10.61}$$

where $r + s + t = 1$, and $(r, s, t) \in [0, 1]$.
   Equation (10.61) can be written:

$$\mathbf{p} = \begin{bmatrix} \mathbf{p}_1 & \mathbf{p}_2 & \mathbf{p}_3 \end{bmatrix} \begin{bmatrix} r \\ s \\ t \end{bmatrix}$$

which is solved using Cramer's Rule with

$$r = \frac{|\mathbf{p} \quad \mathbf{p}_2 \quad \mathbf{p}_3|}{|\mathbf{p}_1 \quad \mathbf{p}_2 \quad \mathbf{p}_3|}, \quad s = \frac{|\mathbf{p}_1 \quad \mathbf{p} \quad \mathbf{p}_3|}{|\mathbf{p}_1 \quad \mathbf{p}_2 \quad \mathbf{p}_3|}, \quad t = \frac{|\mathbf{p}_1 \quad \mathbf{p}_2 \quad \mathbf{p}|}{|\mathbf{p}_1 \quad \mathbf{p}_2 \quad \mathbf{p}_3|} \tag{10.62}$$

Fig. 10.22   A point inside a triangle

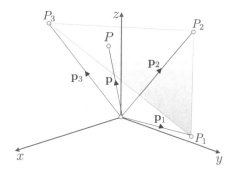

Fig. 10.23 A point inside a triangle

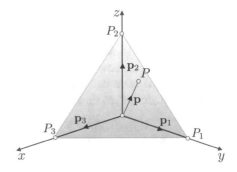

or explicitly:

$$r = \frac{\begin{vmatrix} x_p & x_2 & x_3 \\ y_p & y_2 & y_3 \\ z_p & z_2 & z_3 \end{vmatrix}}{DET}, \qquad s = \frac{\begin{vmatrix} x_1 & x_p & x_3 \\ y_1 & y_p & y_3 \\ z_1 & z_p & z_3 \end{vmatrix}}{DET}, \qquad t = \frac{\begin{vmatrix} x_1 & x_2 & x_p \\ y_1 & y_2 & y_p \\ z_1 & z_2 & z_p \end{vmatrix}}{DET}.$$

where,

$$DET = \begin{vmatrix} x_1 & x_2 & x_3 \\ y_1 & y_2 & y_3 \\ z_1 & z_2 & z_3 \end{vmatrix}.$$

If r, s and t satisfy the above constraints, then P is inside the triangle.

Let's test (10.62) with the scenario shown in Fig. 10.23 with the following points and vectors:

- $P_1(0, 2, 0)$ is a vertex of the triangle,
- $\mathbf{p}_1 = 0\mathbf{i} + 2\mathbf{j} + 0\mathbf{k}$ is P_1's position vector,
- $P_2(0, 0, 2)$ is a vertex of the triangle,
- $\mathbf{p}_2 = 0\mathbf{i} + 0\mathbf{j} + 2\mathbf{k}$ is P_2's position vector,
- $P_3(2, 0, 0)$ is a vertex of the triangle,
- $\mathbf{p}_3 = 2\mathbf{i} + 0\mathbf{j} + 0\mathbf{k}$ is P_3's position vector,
- $P(1, 1, 1)$ is a test point,
- $\mathbf{p} = 1\mathbf{i} + 1\mathbf{j} + 1\mathbf{k}$ is P's position vector.

$$DET = \begin{vmatrix} x_1 & x_2 & x_3 \\ y_1 & y_2 & y_3 \\ z_1 & z_2 & z_3 \end{vmatrix} = \begin{vmatrix} 0 & 0 & 2 \\ 2 & 0 & 0 \\ 0 & 2 & 0 \end{vmatrix} = 8$$

$$r = \frac{\begin{vmatrix} x_p & x_2 & x_3 \\ y_p & y_2 & y_3 \\ z_p & z_2 & z_3 \end{vmatrix}}{DET} = \frac{\begin{vmatrix} 1 & 0 & 2 \\ 1 & 0 & 0 \\ 1 & 2 & 0 \end{vmatrix}}{8} = \frac{1}{2}$$

$$s = \frac{\begin{vmatrix} x_1 & x_p & x_3 \\ y_1 & y_p & y_3 \\ z_1 & z_p & z_3 \end{vmatrix}}{DET} = \frac{\begin{vmatrix} 0 & 1 & 2 \\ 2 & 1 & 0 \\ 0 & 1 & 0 \end{vmatrix}}{8} = \frac{1}{2}$$

$$t = \frac{\begin{vmatrix} x_1 & x_2 & x_p \\ y_1 & y_2 & y_p \\ z_1 & z_2 & z_p \end{vmatrix}}{DET} = \frac{\begin{vmatrix} 0 & 0 & 1 \\ 2 & 0 & 1 \\ 0 & 2 & 1 \end{vmatrix}}{8} = \frac{1}{2}.$$

As $r + s + t > 1$, the point $P(1, 1, 1)$ is not inside the triangle.

Now let's try again with $P\left(\frac{2}{3}, \frac{2}{3}, \frac{2}{3}\right)$:

$$DET = \begin{vmatrix} x_1 & x_2 & x_3 \\ y_1 & y_2 & y_3 \\ z_1 & z_2 & z_3 \end{vmatrix} = \begin{vmatrix} 0 & 0 & 2 \\ 2 & 0 & 0 \\ 0 & 2 & 0 \end{vmatrix} = 8$$

$$r = \frac{\begin{vmatrix} x_p & x_2 & x_3 \\ y_p & y_2 & y_3 \\ z_p & z_2 & z_3 \end{vmatrix}}{DET} = \frac{\begin{vmatrix} \frac{2}{3} & 0 & 2 \\ \frac{2}{3} & 0 & 0 \\ \frac{2}{3} & 2 & 0 \end{vmatrix}}{8} = \frac{1}{3}$$

$$s = \frac{\begin{vmatrix} x_1 & x_p & x_3 \\ y_1 & y_p & y_3 \\ z_1 & z_p & z_3 \end{vmatrix}}{DET} = \frac{\begin{vmatrix} 0 & \frac{2}{3} & 2 \\ 2 & \frac{2}{3} & 0 \\ 0 & \frac{2}{3} & 0 \end{vmatrix}}{8} = \frac{1}{3}$$

$$t = \frac{\begin{vmatrix} x_1 & x_2 & x_p \\ y_1 & y_2 & y_p \\ z_1 & z_2 & z_p \end{vmatrix}}{DET} = \frac{\begin{vmatrix} 0 & 0 & \frac{2}{3} \\ 2 & 0 & \frac{2}{3} \\ 0 & 2 & \frac{2}{3} \end{vmatrix}}{8} = \frac{1}{3}.$$

As $r + s + t = 1$, the point $P\left(\frac{2}{3}, \frac{2}{3}, \frac{2}{3}\right)$ is inside the triangle.

Lastly, let's try with $P(2, 0, 0)$:

$$DET = \begin{vmatrix} x_1 & x_2 & x_3 \\ y_1 & y_2 & y_3 \\ z_1 & z_2 & z_3 \end{vmatrix} = \begin{vmatrix} 0 & 0 & 2 \\ 2 & 0 & 0 \\ 0 & 2 & 0 \end{vmatrix} = 8$$

$$r = \frac{\begin{vmatrix} x_p & x_2 & x_3 \\ y_p & y_2 & y_3 \\ z_p & z_2 & z_3 \end{vmatrix}}{DET} = \frac{\begin{vmatrix} 2 & 0 & 2 \\ 0 & 0 & 0 \\ 0 & 2 & 0 \end{vmatrix}}{8} = 0$$

$$s = \frac{\begin{vmatrix} x_1 & x_p & x_3 \\ y_1 & y_p & y_3 \\ z_1 & z_p & z_3 \end{vmatrix}}{DET} = \frac{\begin{vmatrix} 0 & 2 & 2 \\ 2 & 0 & 0 \\ 0 & 0 & 0 \end{vmatrix}}{8} = 0$$

$$t = \frac{\begin{vmatrix} x_1 & x_2 & x_p \\ y_1 & y_2 & y_p \\ z_1 & z_2 & z_p \end{vmatrix}}{DET} = \frac{\begin{vmatrix} 0 & 0 & 2 \\ 2 & 0 & 0 \\ 0 & 2 & 0 \end{vmatrix}}{8} = 1.$$

As $r + s + t = 1$, the test point is classified as inside, and as two of the barycentric coordinates are zero, it must be the vertex: $P_3(2, 0, 0)$.

10.13   A Sphere Intersecting a Plane

Detecting collisions between irregular objects is difficult. However, it is greatly simplified by enclosing an object within a tight-fitting sphere and calculating collisions between spheres. As part of this analysis let's investigate the geometric relationship between a sphere and a plane. But apart from finding the conditions for collision, let's also find the diameter of the circle when a sphere intersects a plane.

We begin by positioning a sphere juxtaposed with a plane as shown in Fig. 10.24, with the following points and vectors:

- $P(x, y, z)$ is a point of the plane,
- $\mathbf{p} = x\mathbf{i} + y\mathbf{j} + z\mathbf{k}$ is P's position vector,
- $ax + by + cz = d$ is the plane's equation,
- $\hat{\mathbf{n}} = a\mathbf{i} + b\mathbf{j} + c\mathbf{k}$ is the plane's unit vector,
- d is the perpendicular distance from the origin to the plane,
- $C(x_c, y_c, z_c)$ is the centre of the sphere,
- $\mathbf{c} = x_c\mathbf{i} + y_c\mathbf{j} + z_c\mathbf{k}$ is C's position vector,
- r is the sphere's radius,
- $\overrightarrow{CP}$ is the vector between the centre C and P.

The shortest line between the sphere's centre C and the plane is a perpendicular. Naturally, this line is parallel with the plane's normal $\hat{\mathbf{n}}$, and is represented by $\overrightarrow{CP}$. The length of $\overrightarrow{CP}$: $\|\overrightarrow{CP}\|$, determines whether the sphere intersects, touches or misses

Fig. 10.24 A sphere and a plane

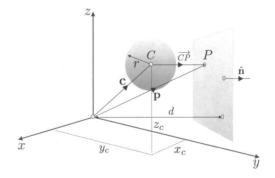

the plane:

$$\|\overrightarrow{CP}\| < r, \quad \text{intersect condition,}$$
$$\|\overrightarrow{CP}\| = r, \quad \text{touch condition,}$$
$$\|\overrightarrow{CP}\| > r, \quad \text{miss condition.}$$

As $\|\hat{\mathbf{n}}\| = 1$, d represents the perpendicular distance from the origin to the plane. Therefore, we can state that

$$\overrightarrow{CP} = \lambda\hat{\mathbf{n}}. \tag{10.63}$$

The position vector $\mathbf{p}$ is defined as:

$$\mathbf{p} = \mathbf{c} + \overrightarrow{CP} = \mathbf{c} + \lambda\hat{\mathbf{n}}. \tag{10.64}$$

Knowing $\hat{\mathbf{n}} \cdot \mathbf{p} = d$, multiply (10.64) by $\hat{\mathbf{n}}$ using the dot product:

$$\hat{\mathbf{n}} \cdot \mathbf{p} = \hat{\mathbf{n}} \cdot \mathbf{c} + \lambda\hat{\mathbf{n}} \cdot \hat{\mathbf{n}}$$
$$d = \hat{\mathbf{n}} \cdot \mathbf{c} + \lambda$$
$$\lambda = d - \hat{\mathbf{n}} \cdot \mathbf{c}. \tag{10.65}$$

But from (10.63), we see that:

$$\|\overrightarrow{CP}\| = |\lambda|$$

and by finding $|\lambda|$, we find $\|\overrightarrow{CP}\|$, and the problem is solved.

Let's illustrate this technique with the following example: a plane is defined by

$$\frac{\sqrt{3}}{3}x + \frac{\sqrt{3}}{3}y + \frac{\sqrt{3}}{3}z = 10$$

and as its normal vector is a unit vector, the perpendicular distance from the origin to the plane is 10, as shown in Fig. 10.25.

Fig. 10.25 A sphere and a plane

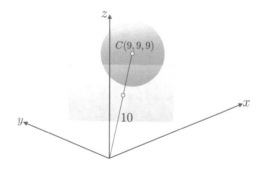

Fig. 10.26 A sphere and a
plane

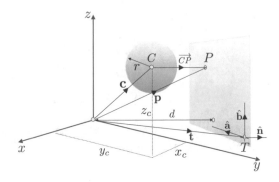

The sphere's radius is 5, and its centre is located at $C(9, 9, 9)$ which makes it $\sqrt{243} \approx 15.588$ from the origin. Obviously, in this example we know that the sphere and plane do not intersect.

Using (10.65):

$$|\lambda| = |d - \hat{n} \cdot \mathbf{c}|$$

where,

$$d = 10, \quad \hat{n} = \frac{\sqrt{3}}{3}\mathbf{i} + \frac{\sqrt{3}}{3}\mathbf{j} + \frac{\sqrt{3}}{3}\mathbf{k}, \quad \mathbf{c} = 9\mathbf{i} + 9\mathbf{j} + 9\mathbf{k}.$$

Then:

$$|\lambda| = \left| 10 - 9\frac{\sqrt{3}}{3} - 9\frac{\sqrt{3}}{3} - 9\frac{\sqrt{3}}{3} \right| \approx |-5.588|$$

$$|\lambda| \approx 5.588$$

and the sphere misses the plane because $|\lambda| > 5$.

If the plane is represented parametrically, as shown in Fig. 10.26, then using two orthogonal, unit vectors $\hat{\mathbf{a}}$ and $\hat{\mathbf{b}}$ gives:

$$\mathbf{p} = \mathbf{t} + \alpha\hat{\mathbf{a}} + \beta\hat{\mathbf{b}}$$

where α and β are scalars, we can proceed as follows.

We find the plane's normal vector $\hat{\mathbf{n}}$ using the cross product:

$$\hat{\mathbf{n}} = \hat{\mathbf{a}} \times \hat{\mathbf{b}}.$$

Therefore,

$$\hat{\mathbf{n}} \cdot \mathbf{t} = d$$

Fig. 10.27 A sphere and a
parametric plane

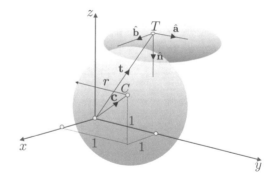

which can be substituted in (10.65):

$$\lambda = \hat{\mathbf{n}} \cdot \mathbf{t} - \hat{\mathbf{n}} \cdot \mathbf{c}$$
$$= \hat{\mathbf{n}} \cdot (\mathbf{t} - \mathbf{c})$$

and

$$|\lambda| = \left| \hat{\mathbf{n}} \cdot (\mathbf{t} - \mathbf{c}) \right|.$$

Let's test this technique with a simple example shown in Fig. 10.27, and as the plane
is placed 2 units above the ground plane, which is the sphere's diameter, they touch.
The plane's equation is

$$\mathbf{p} = \mathbf{t} + \alpha \hat{\mathbf{a}} + \beta \hat{\mathbf{b}}$$

where,

$$\hat{\mathbf{a}} = \mathbf{j}, \quad \hat{\mathbf{b}} = \mathbf{i}, \quad \mathbf{t} = 1\mathbf{j} + 2\mathbf{k}, \quad \mathbf{c} = 1\mathbf{i} + 1\mathbf{j} + 1\mathbf{k}.$$

Calculate $\hat{\mathbf{n}}$:

$$\hat{\mathbf{n}} = \hat{\mathbf{a}} \times \hat{\mathbf{b}} = \begin{vmatrix} \mathbf{i} & \mathbf{j} & \mathbf{k} \\ 0 & 1 & 0 \\ 1 & 0 & 0 \end{vmatrix} = 0\mathbf{i} + 0\mathbf{j} - 1\mathbf{k}$$

and

$$\hat{\mathbf{n}} = 0\mathbf{i} + 0\mathbf{j} - 1\mathbf{k}.$$

Therefore,

$$|\lambda| = |(0\mathbf{i} + 0\mathbf{j} - 1\mathbf{k}) \cdot ((0\mathbf{i} + 1\mathbf{j} + 2\mathbf{k}) - (1\mathbf{i} + 1\mathbf{j} + 1\mathbf{k})|$$
$$= |(0\mathbf{i} + 0\mathbf{j} - 1\mathbf{k}) \cdot (-1\mathbf{i} + 0\mathbf{j} + 1\mathbf{k})|$$
$$= 1.$$

As $|\lambda| = r = 1$, the sphere touches the plane.

Fig. 10.28 Geometry for
circle of intersection

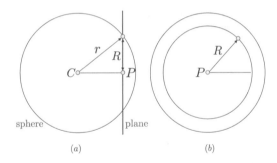

(a)   (b)

The point of contact is given by (10.64):

$$\mathbf{p} = \mathbf{c} + \lambda\hat{\mathbf{n}} = (1\mathbf{i} + 1\mathbf{j} + 1\mathbf{k}) - 1(0\mathbf{i} + 0\mathbf{j} - 1\mathbf{k}) = 1\mathbf{i} + 1\mathbf{j} + 2\mathbf{k}$$

which makes the point of contact $P(1, 1, 2)$.

Now let's calculate the diameter of the circle of intersection when the sphere intersects the plane. Figure 10.28a shows a side view of the sphere and plane, whilst 10.28b shows the intersecting circle. The figures employ the following points and vectors:

- $P(x, y, z)$ is a point of the plane,
- $C(x_c, y_c, z_c)$ is the centre of the sphere,
- r is the sphere's radius,
- $\overrightarrow{CP}$ is the vector between C and P,
- R is the radius of the intersecting circle.

Using the theorem of Pythagoras:

$$r^2 = R^2 + \|\overrightarrow{CP}\|^2 = R^2 + |\lambda|^2.$$

Therefore,

$$R = \sqrt{r^2 - \lambda^2}. \qquad (10.66)$$

10.14 A Sphere Touching a Triangle

Having seen how to detect a sphere touching or intersecting a plane, let's investigate the problem of a sphere touching a triangle. Basically, we need to follow three steps:

1. Derive the triangle's plane equation, probably using the cross product of two vectors taken from the triangle's edges.
2. Determine whether the sphere touches the triangle's plane.
3. If a touch condition occurs, discover if the touch point is inside the triangle.

Fig. 10.29 A sphere and a triangle

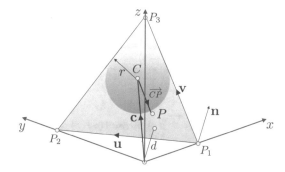

To illustrate how the above steps are implemented, let's investigate a simple example. Figure 10.29 shows a triangle $\Delta P_1 P_2 P_3$ located near the origin and a sphere of radius $r = 1$ with centre $C(1, 1, 2)$.

- $P(x, y, z)$ is a point of the plane,
- $C(1, 1, 2)$ is the centre of the sphere,
- $\mathbf{c} = 1\mathbf{i} + 1\mathbf{j} + 2\mathbf{k}$ is C's position vector,
- $r = 1$ is the sphere's radius,
- $\overrightarrow{CP}$ is the vector between C and P.

We extract two vectors from the triangle:

$$\mathbf{u} = \overrightarrow{P_1 P_2}, \quad \text{and} \quad \mathbf{v} = \overrightarrow{P_1 P_3}$$

where,

$$P_1 = (1, 0, 0), \quad P_2 = (0, 2, 0), \quad P_3 = (0, 0, 3).$$

Therefore,

$$\mathbf{u} = -1\mathbf{i} + 2\mathbf{j} + 0\mathbf{k}, \quad \text{and} \quad \mathbf{v} = -1\mathbf{i} + 0\mathbf{j} + 3\mathbf{k}.$$

Calculate $\mathbf{n}$:

$$\mathbf{n} = \mathbf{u} \times \mathbf{v} = \begin{vmatrix} \mathbf{i} & \mathbf{j} & \mathbf{k} \\ -1 & 2 & 0 \\ -1 & 0 & 3 \end{vmatrix} = 6\mathbf{i} + 3\mathbf{j} + 2\mathbf{k}$$

$$\hat{\mathbf{n}} = \tfrac{6}{7}\mathbf{i} + \tfrac{3}{7}\mathbf{j} + \tfrac{2}{7}\mathbf{k}$$

and

$$\tfrac{6}{7}x + \tfrac{3}{7}y + \tfrac{2}{7}z = d \tag{10.67}$$

where d is the perpendicular distance from the origin to the plane. We can find d by substituting P_1 in (10.67):

$$d = \tfrac{6}{7}.$$

Using (10.65) where $|\lambda| = |d - \hat{\mathbf{n}} \cdot \mathbf{c}|$, we have:

$$|\lambda| = \left| \tfrac{6}{7} - \left(\tfrac{6}{7}\mathbf{i} + \tfrac{3}{7}\mathbf{j} + \tfrac{2}{7}\mathbf{k} \right) \cdot (1\mathbf{i} + 1\mathbf{j} + 2\mathbf{k}) \right|$$
$$= \left| \tfrac{6}{7} - \tfrac{13}{7} \right| = |-1|$$
$$|\lambda| = 1.$$

Remember that $\|\overrightarrow{CP}\| = |\lambda|$, and as $r = 1$, it is a touch condition. The negative value of λ informs us that $\overrightarrow{CP}$ is in the opposite the direction of $\hat{\mathbf{n}}$.

From (10.64):

$$\mathbf{p} = \mathbf{c} + \lambda \hat{\mathbf{n}}$$
$$= (1\mathbf{i} + 1\mathbf{j} + 2\mathbf{k}) - \left(\tfrac{6}{7}\mathbf{i} + \tfrac{3}{7}\mathbf{j} + \tfrac{2}{7}\mathbf{k} \right)$$
$$= \tfrac{1}{7}\mathbf{i} + \tfrac{4}{7}\mathbf{j} + \tfrac{12}{7}\mathbf{k}$$

and

$$P = \left(\tfrac{1}{7}, \tfrac{4}{7}, \tfrac{12}{7} \right).$$

Using Eq. (6.33) we can confirm whether this point is inside the triangle:

$$DET = |\; \mathbf{p}_1 \quad \mathbf{p}_2 \quad \mathbf{p}_3 \;| = \begin{vmatrix} 1 & 0 & 0 \\ 0 & 2 & 0 \\ 0 & 0 & 3 \end{vmatrix} = 6$$

$$r = \frac{|\; \mathbf{p} \quad \mathbf{p}_2 \quad \mathbf{p}_3 \;|}{DET} = \frac{\begin{vmatrix} \tfrac{1}{7} & 0 & 0 \\ \tfrac{4}{7} & 2 & 0 \\ \tfrac{12}{7} & 0 & 3 \end{vmatrix}}{6} = \frac{\tfrac{6}{7}}{6} = \tfrac{1}{7}$$

$$s = \frac{|\; \mathbf{p}_1 \quad \mathbf{p} \quad \mathbf{p}_3 \;|}{DET} = \frac{\begin{vmatrix} 1 & \tfrac{1}{7} & 0 \\ 0 & \tfrac{4}{7} & 0 \\ 0 & \tfrac{12}{7} & 3 \end{vmatrix}}{6} = \frac{\tfrac{12}{7}}{6} = \tfrac{2}{7}$$

$$t = \frac{|\; \mathbf{p}_1 \quad \mathbf{p} \quad \mathbf{p}_3 \;|}{DET} = \frac{\begin{vmatrix} 1 & 0 & \tfrac{1}{7} \\ 0 & 2 & \tfrac{4}{7} \\ 0 & 0 & \tfrac{12}{7} \end{vmatrix}}{6} = \frac{\tfrac{24}{7}}{6} = \tfrac{4}{7}.$$

Note that $r + s + t = 1$, and that they are all positive and less than 1, which means that the point $\left(\tfrac{1}{7}, \tfrac{4}{7}, \tfrac{12}{7} \right)$ is inside the triangle, and the sphere touches the triangle within its boundary.

10.15 Two Intersecting Planes

Hopefully, it is obvious that two planes give rise to a straight line at their intersection. We can discover the vector representing this line by exploiting the fact that the intersecting line is perpendicular to the normal vectors associated with the planes. Therefore, the cross-product of the plane's normal vectors reveals the vector, the direction, of which, is determined by the order of the normal vectors. Figure 10.30 illustrates such a scenario, with the following points and vectors:

- $\mathbf{v} = x_v\mathbf{i} + y_v\mathbf{j} + z_v\mathbf{k}$ is a vector parallel with the line of intersection,
- $P(x, y, z)$ is any point on $\mathbf{v}$,
- $\mathbf{p} = x\mathbf{i} + y\mathbf{j} + z\mathbf{k}$ is P's position vector,
- $T(x_t, y_t, z_t)$ is a point on $\mathbf{v}$ such that $\mathbf{t}$ is orthogonal to $\mathbf{v}$,
- $\mathbf{t} = x_t\mathbf{i} + y_t\mathbf{j} + z_t\mathbf{k}$ is T's position vector,
- $\mathbf{n}_1$ is the first plane's normal vector,
- $\mathbf{n}_2$ is the second plane's normal vector.

To derive a parametric line equation for the intersection we require the coordinates of a point on the line. It is convenient to select a point whose position vector $\mathbf{t}$ is perpendicular to $\mathbf{v}$.

We begin by defining the two plane equations

$$a_1x + b_1y + c_1z = d_1$$
$$a_2x + b_2y + c_2z = d_2$$

where,

$$\mathbf{n}_1 = a_1\mathbf{i} + b_1\mathbf{j} + c_1\mathbf{k}$$
$$\mathbf{n}_2 = a_2\mathbf{i} + b_2\mathbf{j} + c_2\mathbf{k}.$$

Fig. 10.30 Intersecting planes

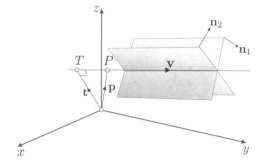

Therefore,

$$\mathbf{n}_1 \times \mathbf{n}_2 = \mathbf{v}$$

where $\mathbf{v}$ is the vector representing the line of intersection.
    The line equation is

$$\mathbf{p} = \mathbf{t} + \lambda \mathbf{v}.$$

As T must satisfy both plane equations:

$$d_1 = \mathbf{n}_1 \cdot \mathbf{t} \qquad\qquad (10.68)$$
$$d_2 = \mathbf{n}_2 \cdot \mathbf{t}. \qquad\qquad (10.69)$$

Furthermore, as $\mathbf{v}$ and $\mathbf{t}$ are orthogonal,

$$0 = \mathbf{v} \cdot \mathbf{t}. \qquad\qquad (10.70)$$

Equations (10.68), (10.69), (10.70) can be combined and written as

$$\begin{bmatrix} d_1 \\ d_2 \\ 0 \end{bmatrix} = \begin{bmatrix} a_1 & b_1 & c_1 \\ a_2 & b_2 & c_2 \\ a_v & b_v & c_v \end{bmatrix} \begin{bmatrix} x_t \\ y_t \\ z_t \end{bmatrix}$$

and solved using Cramer's Rule, where

$$DET = \begin{vmatrix} a_1 & b_1 & c_1 \\ a_2 & b_2 & c_2 \\ a_v & b_v & c_v \end{vmatrix}$$

$$x_t = \frac{\begin{vmatrix} d_1 & b_1 & c_1 \\ d_2 & b_2 & c_2 \\ 0 & b_v & c_v \end{vmatrix}}{DET}, \quad y_t = \frac{\begin{vmatrix} a_1 & d_1 & c_1 \\ a_2 & d_2 & c_2 \\ a_v & 0 & c_v \end{vmatrix}}{DET}, \quad z_t = \frac{\begin{vmatrix} a_1 & b_1 & d_1 \\ a_2 & b_2 & d_2 \\ a_v & b_v & 0 \end{vmatrix}}{DET}.$$

To illustrate this technique, let's consider a scenario whose outcome can be predicted as shown in Fig. 10.31 with the following points and vectors:

- $\mathbf{v} = x_v \mathbf{i} + y_v \mathbf{j} + z_v \mathbf{k}$ is a vector parallel with the line of intersection,
- $T(x_t, y_t, z_t)$ is a point on $\mathbf{v}$ such that $\mathbf{t}$ is orthogonal to $\mathbf{v}$,
- $\mathbf{t} = x_t \mathbf{i} + y_t \mathbf{j} + z_t \mathbf{k}$ is T's position vector,
- $\mathbf{n}_1$ is the first plane's normal vector,
- $\mathbf{n}_2$ is the second plane's normal vector.

The plane equations are

$$y - z = 1, \quad \text{and} \quad y + z = 1$$

Fig. 10.31 Intersecting
planes

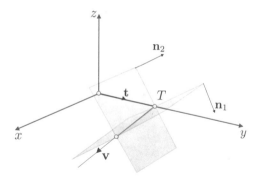

and

$$\mathbf{n}_1 = 0\mathbf{i} + 1\mathbf{j} - 1\mathbf{k}, \quad \text{and} \quad \mathbf{n}_2 = 0\mathbf{i} + 1\mathbf{j} + 1\mathbf{k}.$$

Therefore,

$$\mathbf{v} = \mathbf{n}_1 \times \mathbf{n}_2 = \begin{vmatrix} \mathbf{i} & \mathbf{j} & \mathbf{k} \\ 0 & 1 & -1 \\ 0 & 1 & 1 \end{vmatrix} = 2\mathbf{i} + 0\mathbf{j} + 0\mathbf{k}$$

$$DET = \begin{vmatrix} 0 & 1 & -1 \\ 0 & 1 & 1 \\ 2 & 0 & 0 \end{vmatrix} = 4$$

$$x_t = \frac{1}{4} \begin{vmatrix} 1 & 1 & -1 \\ 1 & 1 & 1 \\ 0 & 0 & 0 \end{vmatrix} = 0$$

$$y_t = \frac{1}{4} \begin{vmatrix} 0 & 1 & -1 \\ 0 & 1 & 1 \\ 2 & 0 & 0 \end{vmatrix} = \frac{4}{4} = 1$$

$$z_t = \frac{1}{4} \begin{vmatrix} 0 & 1 & 1 \\ 0 & 1 & 1 \\ 2 & 0 & 0 \end{vmatrix} = 0.$$

Therefore, the line equation is

$$\mathbf{p} = \mathbf{j} + \lambda\mathbf{i}.$$

Remember that the direction of $\mathbf{v}$ is reversed if $\mathbf{n}_1$ and $\mathbf{n}_2$ are reversed in the cross-product calculation.

10.16 Summary

The last example concludes this chapter on intersections. Hopefully, the reader will feel confident to undertake other intersection problems by implementing some of the above strategies.

Reference

1. Vince J (2017) Mathematics for computer graphics, 5th edn. Springer. ISBN 978-1-4471-7334-2

Chapter 11
Rotating Vectors

11.1 Introduction

This chapter covers the rotation of 2-D vectors about the origin, and 3-D vectors about an axis. It draws upon material I have written about in other books, which you may wish to explore for further information [1–4].

11.2 Rotating Vectors in $\mathbb{R}^2$ About the Origin

We can approach the problem of rotating a 2-D point about the origin in two ways: using the associated coordinate geometry to develop a formula, or developing a matrix that acts upon the point's position vector. Let's choose the second approach, as it is simpler.

We start by declaring a 2×2 matrix $\mathbf{R}_\theta$, which rotates a vector $\mathbf{p}$, by an angle θ, about the origin:

$$\mathbf{p}' = \mathbf{R}_\theta \mathbf{p}$$

where $\mathbf{p}'$ is the rotated vector.

When $\theta = 0$, $\mathbf{R}_\theta$ is the identity matrix:

$$\mathbf{R}_\theta = \begin{bmatrix} 1 & 0 \\ 0 & 1 \end{bmatrix}.$$

Notice that the first column of $\mathbf{R}_\theta$ is the unit basis vector $\mathbf{i}$, and the second column is the unit basis vector $\mathbf{j}$:

$$\mathbf{i} = \begin{bmatrix} 1 \\ 0 \end{bmatrix}, \quad \mathbf{j} = \begin{bmatrix} 0 \\ 1 \end{bmatrix}.$$

© Springer-Verlag London Ltd., part of Springer Nature 2021
J. Vince, *Vector Analysis for Computer Graphics*,
https://doi.org/10.1007/978-1-4471-7505-6_11

Fig. 11.1 The unit basis
vectors rotated by θ

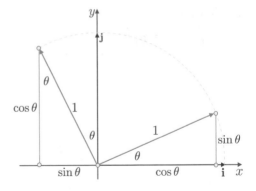

Figure 11.1 shows the unit basis vectors rotated by θ, where vector $\mathbf{i}$ is rotated to $[\cos\theta \quad \sin\theta]$, and vector $\mathbf{j}$ is rotated to $[-\sin\theta \quad \cos\theta]$. Writing these as column vectors, we have:

$$\mathbf{R}_\theta = \begin{bmatrix} \cos\theta & -\sin\theta \\ \sin\theta & \cos\theta \end{bmatrix}. \tag{11.1}$$

For unit scaling, the determinant of R_θ must equal 1:

$$\det \mathbf{R}_\theta = \cos^2\theta + \sin^2\theta = 1.$$

Let's test (11.1) with an example.

Rotate $\mathbf{p} = [1 \quad 0]$ through an angle $90°$:

$$\begin{aligned} \mathbf{p}' &= \begin{bmatrix} \cos 90° & -\sin 90° \\ \sin 90° & \cos 90° \end{bmatrix} \begin{bmatrix} 1 \\ 0 \end{bmatrix} \\ &= \begin{bmatrix} 0 & -1 \\ 1 & 0 \end{bmatrix} \begin{bmatrix} 1 \\ 0 \end{bmatrix} \\ &= \begin{bmatrix} 0 \\ 1 \end{bmatrix}. \end{aligned}$$

Which is correct.

11.3   Rotating Vectors in $\mathbb{R}^3$ About an Axis

In this section we examine three ways of rotating 3-D vectors about an axis: Euler angles, Rodrigues' rotation formula and quaternions.

11.3.1 *Euler Angles*

The Swiss mathematician Leonhard Euler (1707–1783), introduced three angles for rotating a rigid object relative to a fixed coordinate system. Today, they are known as Euler angles, and can readily be expressed as 3×3 matrices. There are numerous ways of describing and combining the angles, but in this short description I will adopt a simple convention that can be exchanged for any personal preference.

The three Euler angles are the angles used to rotate a rigid object about the three axes. For example, one convention is:

- Rotate about the z-axis by α.
- Rotate about the x-axis by β.
- Rotate about the y-axis by γ.

Figure 11.2 shows a view looking along the z-axis towards the xy-plane which contains the coordinates that are affected by the first rotation. Figure 11.3 shows a view looking along the x-axis towards the yz-plane which contains the coordinates that are affected by the first rotation. Figure 11.4 shows a view looking along the y-axis towards the zx-plane which contains the coordinates that are affected by the first rotation.

Fig. 11.2 Rotating by α about the z-axis

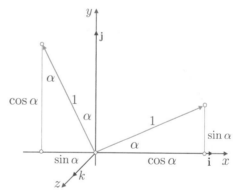

Fig. 11.3 Rotating by β about the x-axis

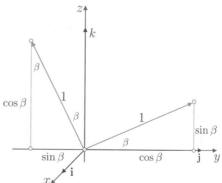

We start by declaring a 3×3 matrix $\mathbf{R}_\alpha$, which rotates a vector $\mathbf{p}$, by an angle α, about the origin:

$$\mathbf{p}' = \mathbf{R}_\alpha \mathbf{p}$$

where $\mathbf{p}'$ is the rotated vector.

When $\alpha = 0$, $\mathbf{R}_\alpha$ is the identity matrix:

$$\mathbf{R}_\alpha = \begin{bmatrix} 1 & 0 & 0 \\ 0 & 1 & 0 \\ 0 & 0 & 1 \end{bmatrix}.$$

Notice that the first column of $\mathbf{R}_\alpha$ is the unit basis vector $\mathbf{i}$, the second column is the unit basis vector $\mathbf{j}$, and the third column is the unit basis vector $\mathbf{k}$:

$$\mathbf{i} = \begin{bmatrix} 1 \\ 0 \\ 0 \end{bmatrix}, \quad \mathbf{j} = \begin{bmatrix} 0 \\ 1 \\ 0 \end{bmatrix}, \quad \mathbf{k} = \begin{bmatrix} 0 \\ 0 \\ 1 \end{bmatrix}.$$

Figure 11.2 shows the unit basis vectors rotated by α, where vector $\mathbf{i}$ is rotated to $[\cos\alpha \quad \sin\alpha \quad 0]$, and vector $\mathbf{j}$ is rotated to $[-\sin\alpha \quad \cos\alpha \quad 0]$. Writing these as column vectors, we have:

$$\mathbf{R}_\alpha = \begin{bmatrix} \cos\alpha & -\sin\alpha & 0 \\ \sin\alpha & \cos\alpha & 0 \\ 0 & 0 & 1 \end{bmatrix}. \tag{11.2}$$

Figure 11.3 shows the unit basis vectors rotated by β, where vector $\mathbf{j}$ is rotated to $[0 \quad \cos\beta \quad \sin\beta]$, and vector $\mathbf{k}$ is rotated to $[0 \quad -\sin\beta \quad \cos\beta]$. Writing these as column vectors, we have:

$$\mathbf{R}_\beta = \begin{bmatrix} 1 & 0 & 0 \\ 0 & \cos\beta & -\sin\beta \\ 0 & \sin\beta & \cos\beta \end{bmatrix}. \tag{11.3}$$

Figure 11.4 shows the unit basis vectors rotated by γ, where vector $\mathbf{k}$ is rotated to $[\sin\gamma \quad 0 \quad \cos\gamma]$, and vector $\mathbf{i}$ is rotated to $[\cos\gamma \quad 0 \quad -\sin\gamma]$. Writing these as column vectors, we have:

$$\mathbf{R}_\gamma = \begin{bmatrix} \cos\gamma & 0 & \sin\gamma \\ 0 & 1 & 0 \\ -\sin\gamma & 0 & \cos\gamma \end{bmatrix}. \tag{11.4}$$

The matrices $\mathbf{R}_\alpha, \mathbf{R}_\beta, \mathbf{R}_\gamma$ can be used individually, or chained together in any sequence. However, when chained together, and the angles are multiples of $90°$, it

Fig. 11.4 Rotating by γ
about the y-axis

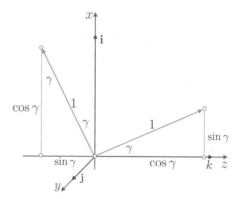

is possible to lose control of a rotational axis. This is known as *gimbal lock*. Let's
show one of the matrices in action.

Rotate the vector $\mathbf{p} = [10 \quad 10 \quad 10]$, $90°$ about the z-axis. This is expressed as
follows:

$$\mathbf{p}' = \mathbf{R}_\alpha \mathbf{p}$$

$$= \begin{bmatrix} \cos 90° & -\sin 90° & 0 \\ \sin 90° & \cos 90° & 0 \\ 0 & 0 & 1 \end{bmatrix} \begin{bmatrix} 10 \\ 10 \\ 10 \end{bmatrix}$$

$$= \begin{bmatrix} 0 & -1 & 0 \\ 1 & 0 & 0 \\ 0 & 0 & 1 \end{bmatrix} \begin{bmatrix} 10 \\ 10 \\ 10 \end{bmatrix}$$

$$= \begin{bmatrix} -10 \\ 10 \\ 10 \end{bmatrix}.$$

Which is correct.

11.3.2 Rodrigues' Rotation Formula

The French banker and mathematician Benjamin Olinde Rodrigues, received his
doctorate in mathematics from the University of Paris, where his dissertation con-
tained a formula showing how a rigid body is rotated about an arbitrary axis. Today,
Rodrigues' rotation formula is expressed:

$$\mathbf{p}' = (1 - \cos\theta)(\hat{\mathbf{v}} \cdot \mathbf{p})\hat{\mathbf{v}} + \cos\theta\mathbf{p} + \sin\theta(\hat{\mathbf{v}} \times \mathbf{p}) \qquad (11.5)$$

where,

$\mathbf{v}$ is the unit vector defining the arbitrary axis,
$\mathbf{p}$ is the vector to be rotated,
θ is the angle of rotation,
$\mathbf{p}'$ is the rotated vector.

Let's illustrate (11.5) with the previous example where vector $\mathbf{p} = [10 \quad 10 \quad 10]$ is rotated 90° about the z-axis. This makes $\hat{\mathbf{v}} = \mathbf{k} = [0 \quad 0 \quad 1]$ and $\theta = 90°$. Therefore, $\cos\theta = 0$ and $\sin\theta = 1$:

$$
\begin{aligned}
\mathbf{p}' &= [0 \quad 0 \quad 1] \times [10 \quad 10 \quad 10] + [0 \quad 0 \quad 1]([0 \quad 0 \quad 1] \cdot [10 \quad 10 \quad 10]) \\
&= [-10 \quad 10 \quad 0] + [0 \quad 0 \quad 1]10 \\
&= [-10 \quad 10 \quad 10].
\end{aligned}
$$

Which is correct.

Rodrigues' rotation formula can also be expressed in matrix notation:

$$\mathbf{R} = \mathbf{I} + (\sin\theta)\mathbf{K} + (1 - \cos\theta)\mathbf{K}^2 \qquad (11.6)$$

where,

$\mathbf{R}$ is the rotation matrix,
$\mathbf{I}$ is the identity matrix,
θ is the angle of rotation,
$\hat{\mathbf{v}} = v_x\mathbf{i} + v_y\mathbf{j} + v_z\mathbf{k}$ is the unit vector defining the arbitrary axis,
$\mathbf{K}$ is the cross-product matrix

$$
\mathbf{K} = \begin{bmatrix} 0 & -v_z & v_y \\ v_z & 0 & -v_x \\ -v_y & v_x & 0 \end{bmatrix}.
$$

Let's illustrate (11.6) with the previous example where vector $\mathbf{p} = [10 \quad 10 \quad 10]$ is rotated 90° about the z-axis. This makes $\hat{\mathbf{v}} = [0 \quad 0 \quad 1]$ and $\theta = 90°$. Therefore, $\cos\theta = 0$ and $\sin\theta = 1$:

$$K = \begin{bmatrix} 0 & -1 & 0 \\ 1 & 0 & 0 \\ 0 & 0 & 0 \end{bmatrix}$$

$$K^2 = \begin{bmatrix} 0 & -1 & 0 \\ 1 & 0 & 0 \\ 0 & 0 & 0 \end{bmatrix} \begin{bmatrix} 0 & -1 & 0 \\ 1 & 0 & 0 \\ 0 & 0 & 0 \end{bmatrix} = \begin{bmatrix} -1 & 0 & 0 \\ 0 & -1 & 0 \\ 0 & 0 & 0 \end{bmatrix}$$

$$R = I + K + K^2$$

$$= \begin{bmatrix} 1 & 0 & 0 \\ 0 & 1 & 0 \\ 0 & 0 & 1 \end{bmatrix} + \begin{bmatrix} 0 & -1 & 0 \\ 1 & 0 & 0 \\ 0 & 0 & 0 \end{bmatrix} + \begin{bmatrix} -1 & 0 & 0 \\ 0 & -1 & 0 \\ 0 & 0 & 0 \end{bmatrix}$$

$$= \begin{bmatrix} 0 & -1 & 0 \\ 1 & 0 & 0 \\ 0 & 0 & 1 \end{bmatrix}.$$

Which is R_α in the original example.

11.3.3 Quaternions

If a point (x, y) is turned into a complex number: $x + yi$, it can be rotated to another point by multiplying it by a second complex number. For example, the point $(1, 1)$ is rotated through $30°$ by multiplying it by $\cos 30° + i \sin 30°$:

$$(1 + i1)\left(\tfrac{\sqrt{3}}{2} + i\tfrac{1}{2} \right) = \tfrac{\sqrt{3}}{2} + i\tfrac{1}{2} + i\tfrac{\sqrt{3}}{2} + i^2\tfrac{1}{2}$$

$$= \tfrac{\sqrt{3}-1}{2} + i\tfrac{\sqrt{3}+1}{2}$$

which makes the new point $\left(\tfrac{\sqrt{3}-1}{2}, \tfrac{\sqrt{3}+1}{2} \right)$.

I doubt whether 3-D rotations were the objective of Hamilton's search for a 3-D equivalent of a complex number, but quaternions do solve this problem. In fact, Hamilton found that sandwiching the point to be rotated between two quaternions resulted in the desired rotation:

$$qpq^{-1}$$

where p encodes the position vector for the point to be rotated, q is the quaternion controlling the axis and angle of rotation, and q^{-1} is q's inverse.

Today, there are two ways of defining a quaternion:

$$q = [s, \ \mathbf{v}] \tag{11.7}$$

$$q = [s + \mathbf{v}]. \tag{11.8}$$

The difference is rather subtle: (11.7) separates the scalar s, and the vector $\mathbf{v}$, with a comma, whereas (11.8) preserves the '+' sign as used in complex numbers. I will employ the comma notation for the rest of this section.

A quaternion q is the combination of a scalar and a vector:

$$q = [s, \ \mathbf{v}]$$

where s is a scalar and $\mathbf{v}$ is a 3-D vector. If we express the quaternion q in terms of its components, we have in an algebraic form:

$$q = [s, \ x\mathbf{i} + y\mathbf{j} + z\mathbf{k}], \quad \text{where} \quad s, x, y, z \in \mathbb{R}.$$

11.3.4   Adding and Subtracting Quaternions

Given two quaternions q_1 and q_2:

$$q_1 = [s_1, \ \mathbf{v}_1] = [s_1, \ x_1\mathbf{i} + y_1\mathbf{j} + z_1\mathbf{k}]$$
$$q_2 = [s_2, \ \mathbf{v}_2] = [s_2, \ x_2\mathbf{i} + y_2\mathbf{j} + z_2\mathbf{k}]$$

they are equal, only if their corresponding terms are equal. Furthermore, like vectors, they can be added and subtracted as follows:

$$q_1 \pm q_2 = [(s_1 \pm s_2), (x_1 \pm x_2)\mathbf{i} + (y_1 \pm y_2)\mathbf{j} + (z_1 \pm z_2)\mathbf{k}].$$

11.3.5   Multiplying Quaternions

When multiplying quaternions we must employ the following rules:

$$\mathbf{i}^2 = \mathbf{j}^2 = \mathbf{k}^2 = \mathbf{ijk} = -1$$
$$\mathbf{ij} = \mathbf{k}, \quad \mathbf{jk} = \mathbf{i}, \quad \mathbf{ki} = \mathbf{j}$$
$$\mathbf{ji} = -\mathbf{k}, \quad \mathbf{kj} = -\mathbf{i}, \quad \mathbf{ik} = -\mathbf{j}.$$

Note that whilst quaternion addition is commutative, the rules make quaternion products non-commutative.

Given two quaternions q_1 and q_2:

$$q_1 = [s_1, \ \mathbf{v}_1] = [s_1, \ x_1\mathbf{i} + y_1\mathbf{j} + z_1\mathbf{k}]$$
$$q_2 = [s_2, \ \mathbf{v}_2] = [s_2, \ x_2\mathbf{i} + y_2\mathbf{j} + z_2\mathbf{k}]$$

the product q_1q_2 is given by:

$$q_1 q_2 = \left[(s_1 s_2 - x_1 x_2 - y_1 y_2 - z_1 z_2), \ (s_1 x_2 + s_2 x_1 + y_1 z_2 - y_2 z_1)\mathbf{i} \right.$$
$$\left. + (s_1 y_2 + s_2 y_1 + z_1 x_2 - z_2 x_1)\mathbf{j} + (s_1 z_2 + s_2 z_1 + x_1 y_2 - x_2 y_1)\mathbf{k} \right]$$

which can be rewritten using the dot and cross product notation as

$$q_1 q_2 = \left[(s_1 s_2 - \mathbf{v}_1 \cdot \mathbf{v}_2), \ s_1 \mathbf{v}_2 + s_2 \mathbf{v}_1 + \mathbf{v}_1 \times \mathbf{v}_2 \right]$$

where,

$$s_1 s_2 - \mathbf{v}_1 \cdot \mathbf{v}_2 \quad \text{is a scalar}$$

and

$$s_1 \mathbf{v}_2 + s_2 \mathbf{v}_1 + \mathbf{v}_1 \times \mathbf{v}_2 \quad \text{is a vector.}$$

11.3.6 Pure Quaternion

A pure quaternion has a zero scalar term:

$$q = [0, \ \mathbf{v}].$$

Therefore, given two pure quaternions:

$$q_1 = [0, \ \mathbf{v}_1] = [0, \ x_1 \mathbf{i} + y_1 \mathbf{j} + z_1 \mathbf{k}]$$
$$q_2 = [0, \ \mathbf{v}_2] = [0, \ x_2 \mathbf{i} + y_2 \mathbf{j} + z_2 \mathbf{k}]$$

their product is another pure quaternion:

$$q_1 q_2 = [0, \ \mathbf{v}_1 \times \mathbf{v}_2].$$

11.3.7 The Inverse Quaternion

Given the quaternion

$$q = [s, \ x\mathbf{i} + y\mathbf{j} + z\mathbf{k}]$$

its inverse q^{-1} is given by:

$$q^{-1} = [s, \ -x\mathbf{i} - y\mathbf{j} - z\mathbf{k}].$$

It can also be shown that

$$q q^{-1} = q^{-1} q = 1.$$

11.3.8   Unit Quaternion

A unit quaternion has a magnitude equal to 1:

$$|q| = \sqrt{s^2 + x^2 + y^2 + z^2} = 1.$$

11.3.9   Rotating Vectors About an Axis

Quaternions are used with vectors rather than individual points. Therefore, in order to manipulate a single point, it is turned into a position vector. A point is then represented in quaternion form by its equivalent position vector with a zero scalar term. For example, the point $P(x, y, z)$ is represented in quaternion form by

$$p = [0, \ x\mathbf{i} + y\mathbf{j} + z\mathbf{k}]$$

which is transformed into another position vector using the process described below. The coordinates of the rotated point are the components of the rotated position vector. This may appear complicated, but in reality it turns out to be rather simple. Let's consider how this is achieved.

Hamilton showed that a pure quaternion p is rotated about an axis using the following operation:

$$p' = qpq^{-1}$$

where the axis and angle of rotation are encoded within the unit quaternion q, and p' stores the rotated vector. For example, to rotate a point $P(x, y, z)$ about an axis $\hat{\mathbf{v}}$, we use the following steps:

1. Convert the point $P(x, y, z)$ to a pure quaternion p:

$$p = [0, \ x\mathbf{i} + y\mathbf{j} + z\mathbf{k}].$$

2. Define the axis of rotation as a unit vector $\hat{\mathbf{v}}$:

$$\hat{\mathbf{v}} = v_x\mathbf{i} + v_y\mathbf{j} + v_z\mathbf{k}$$

3. Define the transforming quaternion q:

$$q = \left[\cos\left(\tfrac{\theta}{2}\right), \ \sin\left(\tfrac{\theta}{2}\right)\hat{\mathbf{v}} \right].$$

4. Define the inverse of the transforming quaternion q^{-1}:

$$q^{-1} = \left[\cos\left(\tfrac{\theta}{2}\right), \ -\sin\left(\tfrac{\theta}{2}\right)\hat{\mathbf{v}}\right].$$

5. Compute p':

$$p' = qpq^{-1}.$$

6. Unpack (x', y', z'):

$$P'(x', y', z') \ \Leftarrow \ p' = \left[0, \ x'\mathbf{i} + y'\mathbf{j} + z'\mathbf{k}\right].$$

We can verify the action of the above transform with a simple example.

Consider the point $P(0, 1, 1)$ in Fig. 11.5, with position vector $\mathbf{p} = 0\mathbf{i} + 1\mathbf{j} + 1\mathbf{k}$, which is to be rotated $90°$ about the y-axis. We can see that the rotated point P' has the coordinates $(1, 1, 0)$ which we will confirm algebraically. The position vector $\mathbf{p}$ is represented by the quaternion p:

$$p = [0, \ 0\mathbf{i} + 1\mathbf{j} + 1\mathbf{k}]$$

and is rotated by evaluating the quaternion p':

$$p' = qpq^{-1}$$

which stores the rotated vector. The axis of rotation is $\mathbf{j}$, therefore the unit quaternion q is given by

$$q = \left[\cos\left(\tfrac{90°}{2}\right), \ \sin\left(\tfrac{90°}{2}\right)(0\mathbf{i} + 1\mathbf{j} + 0\mathbf{k})\right]$$
$$= \left[\cos 45°, \ 0\mathbf{i} + \sin 45°\mathbf{j} + 0\mathbf{k}\right].$$

The inverse quaternion q^{-1} is given by

$$q^{-1} = \left[\cos 45°, \ -0\mathbf{i} - \sin 45°\mathbf{j} - 0\mathbf{k}\right].$$

Let's evaluate qpq^{-1} in two stages: $(qp)q^{-1}$, and for clarity let $\alpha = 45°$:

Fig. 11.5 The point $P(0, 1, 1)$ is rotated to $P'(1, 1, 0)$ using a quaternion coincident with the y-axis

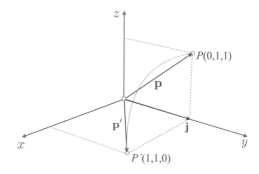

$$(qp) = [\cos\alpha, \ 0\mathbf{i} + \sin\alpha\mathbf{j} + 0\mathbf{k}] \ [0, \ 0\mathbf{i} + 1\mathbf{j} + 0\mathbf{k}]$$
$$= [-\sin\alpha, \ \sin\alpha\mathbf{i} + \cos\alpha\mathbf{j} + \cos\alpha\mathbf{k}]$$
$$(qp)q^{-1} = [-\sin\alpha, \ \sin\alpha\mathbf{i} + \cos\alpha\mathbf{j} + \cos\alpha\mathbf{k}] [\cos\alpha, \ -0\mathbf{i} - \sin\alpha\mathbf{j} - 0\mathbf{k}]$$
$$= [-\sin\alpha\cos\alpha + \sin\alpha\cos\alpha,$$
$$+ \sin^2\alpha\mathbf{j} + \sin\alpha\cos\alpha\mathbf{i} + \cos^2\alpha\mathbf{j} + \cos^2\alpha\mathbf{k}$$
$$+ \begin{vmatrix} \mathbf{i} & \mathbf{j} & \mathbf{k} \\ \sin\alpha & \cos\alpha & \cos\alpha \\ 0 & -\sin\alpha & 0 \end{vmatrix}]$$
$$= [0, \ \sin\alpha\cos\alpha\mathbf{i} + (\sin^2\alpha + \cos^2\alpha)\mathbf{j} + \cos^2\alpha\mathbf{k} + \sin\alpha\cos\alpha\mathbf{i} - \sin^2\alpha\mathbf{k}]$$
$$= [0, \ 2\sin\alpha\cos\alpha\mathbf{i} + (\sin^2\alpha + \cos^2\alpha)\mathbf{j} + (\cos^2\alpha - \sin^2\alpha)\mathbf{k}]$$
$$p' = [0, \ 1\mathbf{i} + 1\mathbf{j} + 0\mathbf{k}].$$

The vector component of p' confirms that P is indeed rotated to $(1, 1, 0)$. Let's evaluate another example before continuing.

Consider a rotation about the z-axis as illustrated in Fig. 11.6. The original point has coordinates $(0, 1, 1)$ and is rotated $-90°$ to $(1, 0, 1)$. This time the quaternion q is defined by

$$q = \left[\cos\left(\frac{-90°}{2}\right), \ \sin\left(\frac{-90°}{2}\right)(0\mathbf{i} + 0\mathbf{j} + 1\mathbf{k})\right]$$
$$= \left[\cos 45°, \ 0\mathbf{i} + 0\mathbf{j} - \sin 45°\mathbf{k}\right]$$

with its inverse

$$q^{-1} = \left[\cos 45°, \ 0\mathbf{i} + 0\mathbf{j} + \sin 45°\mathbf{k}\right]$$

and the point to be rotated in quaternion form is

$$p = [0, \ 0\mathbf{i} + 1\mathbf{j} + 1\mathbf{k}].$$

Fig. 11.6 The point $P(0, 1, 1)$ is rotated $-90°$ to $P'(1, 0, 1)$ using a quaternion coincident with the z-axis

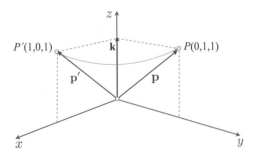

Evaluating this in two stages with $\alpha = 45°$, we have:

$$(qp) = [\cos\alpha, \ 0\mathbf{i} + 0\mathbf{j} - \sin\alpha\mathbf{k}][0, \ 0\mathbf{i} + 1\mathbf{j} + 1\mathbf{k}]$$

$$= \left[\sin\alpha, \ \cos\alpha\mathbf{j} + \cos\alpha\mathbf{k} + \begin{vmatrix} \mathbf{i} & \mathbf{j} & \mathbf{k} \\ 0 & 0 & -\sin\alpha \\ 0 & 1 & 1 \end{vmatrix}\right]$$

$$= [\sin\alpha, \ \sin\alpha\mathbf{i} + \cos\alpha\mathbf{j} + \cos\alpha\mathbf{k}]$$

$$(qp)q^{-1} = [\sin\alpha, \ \sin\alpha\mathbf{i} + \cos\alpha\mathbf{j} + \cos\alpha\mathbf{k}][\cos\alpha, \ 0\mathbf{i} + 0\mathbf{j} + \sin\alpha\mathbf{k}]$$

$$= \big[\sin\alpha\cos\alpha - \sin\alpha\cos\alpha,$$

$$+ \sin^2\alpha\mathbf{k} + \sin\alpha\cos\alpha\mathbf{i} + \cos^2\alpha\mathbf{j} + \cos^2\alpha\mathbf{k}$$

$$+ \begin{vmatrix} \mathbf{i} & \mathbf{j} & \mathbf{k} \\ \sin\alpha\cos\alpha & \cos\alpha & \cos\alpha \\ 0 & 0 & \sin\alpha \end{vmatrix}\big]$$

$$= \big[0, \ \sin\alpha\cos\alpha\mathbf{i} + \cos^2\alpha\mathbf{j} + (\sin^2\alpha + \cos^2\alpha)\mathbf{k}$$

$$+ \sin\alpha\cos\alpha\mathbf{i} - \sin^2\alpha\mathbf{j}\big]$$

$$= \big[0, \ 2\sin\alpha\cos\alpha\mathbf{i} + (\cos^2\alpha - \sin^2\alpha)\mathbf{j} + 1\mathbf{k}\big]$$

$$= [0, \ 1\mathbf{i} + 0\mathbf{j} + 1\mathbf{k}].$$

The vector component of p' confirms that the rotated point is $(1, 0, 1)$.

11.3.10 The Double Angle

Now let's show why it's necessary to halve the angle of rotation. We begin by defining a unit-norm quaternion q:

$$q = [s, \ \lambda\hat{\mathbf{v}}]$$

where $s^2 + \lambda^2 = 1$. The vector to be rotated is encoded as a pure quaternion:

$$p = [0, \ \mathbf{p}]$$

and the inverse quaternion q^{-1} is

$$q^{-1} = [s, \ -\lambda\hat{\mathbf{v}}].$$

Therefore, the product qpq^{-1} is

$$
\begin{aligned}
qpq^{-1} &= [s,\ \lambda\hat{\mathbf{v}}][0,\ \mathbf{p}][s,\ -\lambda\hat{\mathbf{v}}] \\
&= [-\lambda\hat{\mathbf{v}}\cdot\mathbf{p},\ s\mathbf{p}+\lambda\hat{\mathbf{v}}\times\mathbf{p}][s,\ -\lambda\hat{\mathbf{v}}] \\
&= [-\lambda s\hat{\mathbf{v}}\cdot\mathbf{p}+\lambda s\mathbf{p}\cdot\hat{\mathbf{v}}+\lambda^2(\hat{\mathbf{v}}\times\mathbf{p})\cdot\hat{\mathbf{v}}, \\
&\quad\ +\lambda^2(\hat{\mathbf{v}}\cdot\mathbf{p})\hat{\mathbf{v}}+s^2\mathbf{p}+\lambda s\hat{\mathbf{v}}\times\mathbf{p}-\lambda s\mathbf{p}\times\hat{\mathbf{v}}-\lambda^2(\hat{\mathbf{v}}\times\mathbf{p})\times\hat{\mathbf{v}}] \\
&= [\lambda^2(\hat{\mathbf{v}}\times\mathbf{p})\cdot\hat{\mathbf{v}},\ \lambda^2(\hat{\mathbf{v}}\cdot\mathbf{p})\hat{\mathbf{v}}+s^2\mathbf{p}+2\lambda s\hat{\mathbf{v}}\times\mathbf{p}-\lambda^2(\hat{\mathbf{v}}\times\mathbf{p})\times\hat{\mathbf{v}}].
\end{aligned}
$$

Note that
$$
(\hat{\mathbf{v}}\times\mathbf{p})\cdot\hat{\mathbf{v}}=0
$$

and
$$
(\hat{\mathbf{v}}\times\mathbf{p})\times\hat{\mathbf{v}}=(\hat{\mathbf{v}}\cdot\hat{\mathbf{v}})\mathbf{p}-(\mathbf{p}\cdot\hat{\mathbf{v}})\hat{\mathbf{v}}=\mathbf{p}-(\mathbf{p}\cdot\hat{\mathbf{v}})\hat{\mathbf{v}}.
$$

Therefore,
$$
\begin{aligned}
qpq^{-1} &= [0,\ \lambda^2(\hat{\mathbf{v}}\cdot\mathbf{p})\hat{\mathbf{v}}+s^2\mathbf{p}+2\lambda s\hat{\mathbf{v}}\times\mathbf{p}-\lambda^2\mathbf{p}+\lambda^2(\mathbf{p}\cdot\hat{\mathbf{v}})\hat{\mathbf{v}}] \\
&= [0,\ 2\lambda^2(\hat{\mathbf{v}}\cdot\mathbf{p})\hat{\mathbf{v}}+(s^2-\lambda^2)\mathbf{p}+2\lambda s\hat{\mathbf{v}}\times\mathbf{p}]. \tag{11.9}
\end{aligned}
$$

Obviously, this is a pure quaternion as the scalar component is zero. However, it is not obvious where the angle doubling comes from. But look what happens when we make $s=\cos\theta$ and $\lambda=\sin\theta$:

$$
\begin{aligned}
qpq^{-1} &= [0,\ 2\sin^2\theta(\hat{\mathbf{v}}\cdot\mathbf{p})\hat{\mathbf{v}}+(\cos^2\theta-\sin^2\theta)\mathbf{p}+2\sin\theta\cos\theta\hat{\mathbf{v}}\times\mathbf{p}] \\
&= [0,\ (1-\cos(2\theta))(\hat{\mathbf{v}}\cdot\mathbf{p})\hat{\mathbf{v}}+\cos(2\theta)\mathbf{p}+\sin(2\theta)\hat{\mathbf{v}}\times\mathbf{p}].
\end{aligned}
$$

The double-angle trigonometric terms emerge! Now, if we want this product to actually rotate the vector by θ, then we must build this in from the outset by halving θ in q:

$$
q=\left[\cos\left(\tfrac{\theta}{2}\right),\ \sin\left(\tfrac{\theta}{2}\right)\hat{\mathbf{v}}\right] \tag{11.10}
$$

which makes

$$
qpq^{-1}=[0,\ (1-\cos\theta)(\hat{\mathbf{v}}\cdot\mathbf{p})\hat{\mathbf{v}}+\cos\theta\mathbf{p}+\sin\theta\hat{\mathbf{v}}\times\mathbf{p}]. \tag{11.11}
$$

The product qpq^{-1} was discovered by Hamilton who failed to publish the result. Cayley, also discovered the product and published the result in 1845 [5]. However, Altmann notes that "in Cayley's collected papers he concedes priority to Hamilton." [6], which was a nice gesture. However, the person who had recognised the importance of the half-angle parameters in (11.10) before Hamilton and Cayley was Rodrigues—who published a solution that was not seen by Hamilton, but apparently, was seen by Cayley.

11.3.11 Quaternion as a Matrix

In order to represent qpq^{-1} as a matrix, it is convenient to describe the unit-norm quaternion as

$$q = [s, \ \mathbf{v}]$$
$$= [s, \ x\mathbf{i} + y\mathbf{j} + z\mathbf{k}]$$

where

$$s^2 + \|\mathbf{v}\| = 1$$

and the pure quaternion as

$$p = \left[0, \ \mathbf{p}\right]$$
$$= \left[0, \ p_x\mathbf{i} + p_y\mathbf{j} + p_z\mathbf{k}\right].$$

A simple way to compute qpq^{-1} is to use (11.9) and substitute $\|\mathbf{v}\|$ for λ:

$$qpq^{-1} = \left[0, \ 2\lambda^2 \left(\hat{\mathbf{v}} \cdot \mathbf{p}\right) \hat{\mathbf{v}} + \left(s^2 - \lambda^2\right)\mathbf{p} + 2\lambda s \hat{\mathbf{v}} \times \mathbf{p}\right]$$
$$= \left[0, \ 2\|\mathbf{v}\|^2 \left(\hat{\mathbf{v}} \cdot \mathbf{p}\right) \hat{\mathbf{v}} + \left(s^2 - \|\mathbf{v}\|^2\right)\mathbf{p} + 2\|\mathbf{v}\|s \hat{\mathbf{v}} \times \mathbf{p}\right].$$

Next, we substitute $\mathbf{v}$ for $\|\mathbf{v}\|\hat{\mathbf{v}}$:

$$qpq^{-1} = \left[0, \ 2\left(\mathbf{v} \cdot \mathbf{p}\right)\mathbf{v} + \left(s^2 - \|\mathbf{v}\|^2\right)\mathbf{p} + 2s\mathbf{v} \times \mathbf{p}\right].$$

Finally, as we are working with unit-norm quaternions to prevent scaling:

$$s^2 + \|\mathbf{v}\|^2 = 1$$

and

$$s^2 - \|\mathbf{v}\|^2 = 2s^2 - 1.$$

Therefore,

$$qpq^{-1} = \left[0, \ 2(\mathbf{v} \cdot \mathbf{p})\mathbf{v} + \left(2s^2 - 1\right)\mathbf{p} + 2s\mathbf{v} \times \mathbf{p}\right].$$

If we let $p' = qpq^{-1}$, which is a pure quaternion, we have

$$p' = qpq^{-1}$$
$$= \left[0, \ \mathbf{p}'\right]$$
$$= \left[0, \ 2(\mathbf{v} \cdot \mathbf{p})\mathbf{v} + \left(2s^2 - 1\right)\mathbf{p} + 2s\mathbf{v} \times \mathbf{p}\right]$$
$$\mathbf{p}' = 2(\mathbf{v} \cdot \mathbf{p})\mathbf{v} + \left(2s^2 - 1\right)\mathbf{p} + 2s\mathbf{v} \times \mathbf{p}.$$

We are only interested in the rotated vector $\mathbf{p}'$ comprising the three terms $2(\mathbf{v} \cdot \mathbf{p})\mathbf{v}$, $(2s^2 - 1)\mathbf{p}$ and $2s\mathbf{v} \times \mathbf{p}$, which can be represented by three individual matrices and summed together:

$$2(\mathbf{v} \cdot \mathbf{p})\mathbf{v} = 2\left(xx_p + yy_p + zz_p\right)(x\mathbf{i} + y\mathbf{j} + z\mathbf{k})$$

$$= \begin{bmatrix} 2x^2 & 2xy & 2xz \\ 2xy & 2y^2 & 2yz \\ 2xz & 2yz & 2z^2 \end{bmatrix} \begin{bmatrix} p_x \\ p_y \\ p_z \end{bmatrix}.$$

$$\left(2s^2 - 1\right)\mathbf{p} = \left(2s^2 - 1\right)x_p\mathbf{i} + \left(2s^2 - 1\right)y_p\mathbf{j} + \left(2s^2 - 1\right)z_p\mathbf{k}$$

$$= \begin{bmatrix} 2s^2 - 1 & 0 & 0 \\ 0 & 2s^2 - 1 & 0 \\ 0 & 0 & 2s^2 - 1 \end{bmatrix} \begin{bmatrix} p_x \\ p_y \\ p_z \end{bmatrix}.$$

$$2s\mathbf{v} \times \mathbf{p} = 2s\left(\left(yz_p - zy_p\right)\mathbf{i} + \left(zx_p - xz_p\right)\mathbf{j} + \left(xy_p - yx_p\right)\mathbf{k}\right)$$

$$= \begin{bmatrix} 0 & -2sz & 2sy \\ 2sz & 0 & -2sx \\ -2sy & 2sx & 0 \end{bmatrix} \begin{bmatrix} p_x \\ p_y \\ p_z \end{bmatrix}.$$

Adding these matrices together:

$$\mathbf{p}' = \begin{bmatrix} 2\left(s^2 + x^2\right) - 1 & 2\left(xy - sz\right) & 2\left(xz + sy\right) \\ 2\left(xy + sz\right) & 2\left(s^2 + y^2\right) - 1 & 2\left(yz - sx\right) \\ 2\left(xz - sy\right) & 2\left(yz + sx\right) & 2\left(s^2 + z^2\right) - 1 \end{bmatrix} \begin{bmatrix} p_x \\ p_y \\ p_z \end{bmatrix} \qquad (11.12)$$

or

$$\mathbf{p}' = \begin{bmatrix} 1 - 2\left(y^2 + z^2\right) & 2\left(xy - sz\right) & 2\left(xz + sy\right) \\ 2\left(xy + sz\right) & 1 - 2\left(x^2 + z^2\right) & 2\left(yz - sx\right) \\ 2\left(xz - sy\right) & 2\left(yz + sx\right) & 1 - 2\left(x^2 + y^2\right) \end{bmatrix} \begin{bmatrix} p_x \\ p_y \\ p_z \end{bmatrix} \qquad (11.13)$$

where,

$$[0, \mathbf{p}'] = qpq^{-1}.$$

Now let's reverse the product. To compute the vector part of $q^{-1}pq$ all that we have to do is reverse the sign of $2s\mathbf{v} \times \mathbf{p}$:

$$\mathbf{p}' = \begin{bmatrix} 2\left(s^2 + x^2\right) - 1 & 2\left(xy + sz\right) & 2\left(xz - sy\right) \\ 2\left(xy - sz\right) & 2\left(s^2 + y^2\right) - 1 & 2\left(yz + sx\right) \\ 2\left(xz + sy\right) & 2\left(yz - sx\right) & 2\left(s^2 + z^2\right) - 1 \end{bmatrix} \begin{bmatrix} p_x \\ p_y \\ p_z \end{bmatrix} \qquad (11.14)$$

or

$$\mathbf{p}' = \begin{bmatrix} 1 - 2\left(y^2 + z^2\right) & 2\left(xy + sz\right) & 2\left(xz - sy\right) \\ 2\left(xy - sz\right) & 1 - 2\left(x^2 + z^2\right) & 2\left(yz + sx\right) \\ 2\left(xz + sy\right) & 2\left(yz - sx\right) & 1 - 2\left(x^2 + y^2\right) \end{bmatrix} \begin{bmatrix} p_x \\ p_y \\ p_z \end{bmatrix} \quad (11.15)$$

where,

$$\left[0, \ \mathbf{p}'\right] = q^{-1}pq.$$

Note that (11.14) is the transpose of (11.12), and (11.15) is the transpose of (11.13).

Let's test (11.12) using the example illustrated in Fig. 11.5, where $\mathbf{p} = 0\mathbf{i} + 1\mathbf{j} + 1\mathbf{k}$ is rotated 90° about the y-axis. This makes $p_x = 0$, $p_y = 1$, $p_z = 1$. The quaternion q takes the form:

$$q = \left[\cos\left(\tfrac{\theta}{2}\right), \ \sin\left(\tfrac{\theta}{2}\right)\hat{\mathbf{v}}\right]$$

which means that $\theta = 90°$ and $\hat{\mathbf{v}} = 1\mathbf{j}$, therefore,

$$q = \left[\cos 45°, \ \sin 45°\mathbf{j}\right].$$

Consequently,

$$s = \tfrac{\sqrt{2}}{2}, \quad x = 0 \quad y = \tfrac{\sqrt{2}}{2}, \quad z = 0.$$

$$\mathbf{p}' = \begin{bmatrix} 2\left(s^2 + x^2 j\right) - 1 & 2\left(xy - sz\right) & 2\left(xz + sy\right) \\ 2\left(xy + szj\right) & 2\left(s^2 + y^2\right) - 1 & 2\left(yz - sx\right) \\ 2\left(xz - syj\right) & 2\left(yz + sx\right) & 2\left(s^2 + z^2\right) - 1 \end{bmatrix} \begin{bmatrix} p_x \\ p_y \\ p_z \end{bmatrix}$$

$$= \begin{bmatrix} 2\left(\tfrac{1}{2} + 0\right) - 1 & 2\left(0 - 0\right) & 2\left(0 + \tfrac{1}{2}\right) \\ 2\left(0 + 0\right) & 2\left(\tfrac{1}{2} + \tfrac{1}{2}\right) - 1 & 2\left(0 - 0\right) \\ 2\left(0 - \tfrac{1}{2}\right) & 2\left(0 + 0\right) & 2\left(\tfrac{1}{2} + 0\right) - 1 \end{bmatrix} \begin{bmatrix} 0 \\ 1 \\ 1 \end{bmatrix}$$

$$\begin{bmatrix} 1 \\ 1 \\ 0 \end{bmatrix} = \begin{bmatrix} 0 & 0 & 1 \\ 0 & 1 & 0 \\ -1 & 0 & 0 \end{bmatrix} \begin{bmatrix} 0 \\ 1 \\ 1 \end{bmatrix}.$$

Which is correct.

11.4   Summary

I will leave it up you to decide whether you use quaternions or Rodrigue's rotation formula to rotate a vector. After all, it is a question of how fast it takes to code an algorithm, and how fast it takes to compute the result.

References

1. Vince J (2017) Mathematics for computer graphics, 5th edn. Springer, Berlin. ISBN 978-1-4471-7334-2
2. Vince J (2011) Rotation transforms for computer graphics. Springer, Berlin. ISBN 978-0-85729-153-0
3. Vince J (2012) Matrix transforms for computer graphics. Springer, Berlin. ISBN 978-1-4471-4320-8
4. Vince J (2011) Quaternions for computer graphics. Springer, Berlin. ISBN 978-0-85729-759-4
5. Cayley A (1848) The collected mathematical papers, vol I, p 586, note 20
6. Altmann SL (1986) Rotations, quaternions and double groups. Dover Publications, p 16. ISBN-13: 978-0-486-44518-2

Index

© Springer-Verlag London Ltd., part of Springer Nature 2021
J. Vince, *Vector Analysis for Computer Graphics*,
https://doi.org/10.1007/978-1-4471-7505-6

Printed in the United States
by Baker & Taylor Publisher Services